By Kevin Roose

Futureproof

Young Money

The Unlikely Disciple

Futureproof

RANDOM HOUSE
NEW YORK

Futureproof

9 Rules for Humans in the Age of Automation

Kevin Roose

Published in the United States by Random House, an imprint and division of Penguin Random House LLC, New York.

RANDOM HOUSE and the HOUSE colophon are registered trademarks of Penguin Random House LLC.

Library of Congress Cataloging-in-Publication Data
Names: Roose, Kevin, author.
Title: Futureproof: 9 rules for humans in the age of automation / Kevin Roose.
Description: New York: Random House, [2021]
Identifiers: LCCN 2020001669 (print) | LCCN 2020001670 (ebook) | ISBN 9780593133347 (hardcover) | ISBN 9780593133354 (ebook)
Subjects: LCSH: Computers and civilization. | Automation—Social aspects. | Artificial intelligence—Social aspects. | Success in business.
Classification: LCC QA76.9.C66 R635 2020 (print) | LCC QA76.9.C66 (ebook) | DDC 303.48/34—dc23
LC record available at https://lccn.loc.gov/2020001669
LC ebook record available at https://lccn.loc.gov/2020001670

Printed in the United States of America on acid-free paper

randomhousebooks.com

9 8 7 6 5 4 3 2 1

First Edition

Book design by Susan Turner

In memory of Dad

Proceed as way opens.

—Quaker proverb

Contents

Introduction

Recently, I was at a party in San Francisco when a man approached me and introduced himself as the founder of a small AI start-up.

As soon as the founder figured out that I was a technology writer for *The New York Times,* he launched into a pitch for his company, which he said was trying to revolutionize the manufacturing sector using a new AI technique called "deep reinforcement learning."

Modern factories, he explained, were struggling with what is called "production planning"—the complex art of calculating which machines should be making which things on which days. Today, he said, most factories employ humans to look at thick piles of data and customer orders to figure out whether the plastic-molding machines should be making X-Men figurines on Tuesdays and TV remotes on Thursdays, or vice versa. It's one of those dull-but-essential tasks without which modern capitalism would probably grind to a halt, and companies spend billions of dollars a year trying to get it right.

The founder explained that his company's AI could run millions of virtual simulations for any given factory, eventually arriving at the exact sequence of processes that would allow it to produce goods most efficiently. This AI, he said, would allow factories to replace entire teams of human production planners, along with most of the outdated software those people relied on.

"We call it the Boomer Remover," he said.

"The . . . Boomer . . . Remover?" I asked.

"Yeah," he said. "I mean, that's not the official name. But our clients have way too many old, overpaid middle-managers who aren't really necessary anymore. Our platform lets them replace those people."

The founder, who appeared to be a few drinks deep, then told a story about a client who had been looking for a way to get rid of one particular production planner for years, but could never figure out how to fully automate his job away. But mere days after installing his company's software, the client had been able to eliminate the planner's position with no loss of efficiency.

Slightly stunned, I asked the founder if he knew what had happened to the production planner. Was he reassigned within the company? Was he just laid off unceremoniously? Did he know that his bosses had been scheming to replace him with a robot?

The founder chuckled.

"That's not my problem," he said, and headed to the bar for another drink.

I've loved technology since I was a kid, when I spent all my free time building websites and saving up allowance money for new PC parts. And for years, I rolled my eyes whenever some-one suggested that computers would destroy jobs, destabilize society, or usher us into a futuristic dystopia. I was especially dismissive of people who predicted that AI would one day make humans obsolete. Weren't these the same panicky tech-nophobes who warned us that Nintendo games would melt our brains? Didn't their fears always end up being overblown?

Several years ago, when I started as a tech columnist for the *Times,* most of what I heard about AI mirrored my own optimistic views. I met with start-up founders and engineers in Silicon Valley who showed me how new advances in fields like deep learning were helping them build all kinds of world-improving tools: algorithms that could increase farmers' crop yields, software that would help hospitals run more efficiently, self-driving cars that could shuttle us around while we took naps and watched Netflix.

This was the euphoric peak of the AI hype cycle, a time when all of the American tech giants—Google, Facebook, Apple, Amazon, Microsoft—were pouring billions of dollars into developing new AI products and shoving machine learn-ing algorithms into as many of their apps as possible. They wrote blank checks to their AI research teams, and poached professors and grad students out of top computer science de-partments with frankly hilarious job offers. (One professor told me, in hushed tones, that a tech company had just offered one of his colleagues a $1 million annual contract that only re-quired him to work on Fridays.) Everywhere you looked, start-ups were raising gargantuan funding rounds, promising to use AI to revolutionize everything from podcasting to pizza deliv-ery. And the conventional wisdom, at least among my sources, was that these new, AI-based tools would be an unequivocally good thing for society.

But in the past few years, as I've spent more time reporting on AI and automation,* three things have made me rethink my optimism.

First, as I studied the history of technological change, I realized that some of the stories technologists liked to tell— like the narrative that technology had always created more jobs than it destroyed, or that humans and AI would collaborate rather than compete with one another—turned out to be, if not false, then at least radically incomplete. (We'll take a closer look at some of these narratives, and the holes they contain, in Chapter 1.)

Second, as I reported on the effects AI and automation were having in the world, I saw a stark gap between the promises these technologies' creators had made and the actual, real-world experiences of the people using them.

I interviewed users of social media platforms like YouTube and Facebook, who had thought that those platforms' AI-driven recommendation systems would help them find interesting and relevant content, but who had instead been led down rabbit holes filled with misinformation and conspiracy

---

* Quick usage note: In this book, I'm going to use "AI and automation" as a catch-all term for various digital processes that carry out tasks that were previously done by humans. Among computer scientists, "AI" most often refers to a subcategory of automation in which computers are programmed to adapt and learn on their own using techniques like machine learning, and a lot of very smart people get annoyed when you call something "AI" that is really just a static, rule-based algorithm. But this distinction can be fuzzy and mostly lost on the nontechnical reader, so I'll hedge my bets by using both terms whenever possible. Likewise, I will keep my earnest use of "robot"—a term many engineers hate, because it's been tainted by sci-fi movies and can be used to describe everything from droids to dishwashers—to a minimum.

theories. I heard about teachers whose schools had implemented high-tech "personalized learning" systems in hopes of improving student outcomes, but who had found themselves fumbling with broken tablet computers and erratic software. I listened to the complaints of Uber and Lyft drivers who had been lured by the promise of flexible employment, but then found themselves suffering under the thumb of a draconian algorithm that nudged them to work longer hours, punished them for taking breaks, and constantly manipulated their pay.

All of these stories seemed to indicate that AI and automation were working well for some people—namely, the executives and investors who built and profited from the technology—but that they weren't making life better for everyone.

The third, and clearest, sign that something was off came in 2019, when I started hearing snippets of a more honest automation conversation.

This conversation wasn't the rosy, optimistic one playing out on tech conference stages and in glossy business magazine spreads. It was happening privately among elites and engineers, like the start-up founder who told me about his Boomer Remover software. These people had seen the future of AI and automation up close, and they had no illusions about where these technologies were headed. They knew that machines are, or soon will be, capable of replacing humans in a wide range of jobs and activities. Some of them were greedily racing toward fully automating their workforces, their eyes bulging with dollar signs like Looney Tunes characters. Others were more worried about the political backlash mass automation could cause, and wanted to engineer a softer landing for the victims. But they all knew that there *would* be victims. None of them were under the impression that AI and automation will be good for everyone, and nobody was even considering pumping the brakes.

I got my first glimpse of this other automation conversation during the World Economic Forum, an annual conference held in Davos, Switzerland. Davos bills itself as a high-minded confab where global elites gather to discuss the world's most pressing problems, but in reality it's more like the Coachella of capitalism—a beyond-satire boondoggle where plutocrats, politicians, and do-gooder celebrities come to see and be seen. It's the only place in the world where it wouldn't be at all unusual for the CEO of Goldman Sachs, the Japanese prime minister, and will.i.am to sit around chatting about income inequality while eating $37 sandwiches.

My bosses at the *Times* had invited me to cover that year's forum, which was focused on "Globalization 4.0"—the essentially meaningless term Davos types had concocted for the emerging economic era defined by this new, transformative wave of AI and automation technology. Every day, I went to panels with titles like "Shaping a New Market Architecture" and "The Factory of the Future," where powerful executives vowed to build "human-centered AI" that would be great for companies and workers alike.

But at night, after their public events were over, the Davos attendees took off their humanitarian masks and got down to business. At lavish, off-the-record dinners and cocktail parties, I watched them grill tech experts about how AI could help transform their companies into sleek, automated profit machines. They gossiped about which automation products their competitors were using. They struck deals with consultants for "digital transformation" projects, which they hoped would save them millions of dollars by shrinking their reliance on human workers.

I ran into one of those consultants one day. His name is Mohit Joshi, and he's the president of a company called Infosys, an India-based consulting firm that helps big businesses automate their operations. When I asked Joshi how his meetings with executives were going, his eyebrows arched, and he told me that the Davos elite's obsession with automation was even more intense than he—a guy who *literally automates jobs for a living*—had expected.

Once, he said, his clients had wanted to reduce their workforces incrementally, keeping maybe 95 percent of their human workers while automating around the edges.

"But now," he told me, "they're saying, 'Why can't we do it with *one percent* of the people we have?'"

In other words, when the cameras and microphones were off, these executives weren't talking about helping workers. They were fantasizing about getting rid of them completely.

After coming home from Davos, I decided to learn as much about AI and automation as I could. I wanted to know: What was actually happening inside companies and engineering departments? What kinds of people were in danger of being replaced by machines? What, if anything, could we do to protect ourselves?

So, for months, I interviewed engineers, executives, investors, politicians, economists, and historians. I visited research labs and start-ups, and went to tech conferences and industry meet-ups. I read approximately a hundred books whose cover art consists of a robot shaking hands with a human.

As I was reporting, the public conversation around automation began to shed some of its optimistic sheen. People started noticing the destructive effects of social media algorithms, which entrapped users in ideological echo chambers and nudged them toward more extreme beliefs. Tech leaders like Bill Gates and Elon Musk warned that AI could put millions of people out of work and urged politicians to take it seriously as a threat. Economists began making gloomy predictions about what AI would do to workers, and politicians began stumping about the need for radical solutions to fend off an automation-fueled unemployment crisis. The most prominent public figure to sound the alarm, the New York businessman Andrew Yang, ran for the Democratic nomination for president in 2020 on a promise to give all Americans a $1,000-a-month "freedom dividend" to cushion the blow of automation. He didn't win, but his warning of a looming AI revolution entered the zeitgeist and pushed the conversation about technological unemployment into the mainstream.

Fears of job-killing machines aren't new. In fact, they date back to roughly 350 B.C.E., when Aristotle mused that automated weavers and self-playing harps could reduce the demand for slave labor. Since then, machine-related anxieties have ebbed and flowed, often peaking during periods of rapid technological change. In 1928, *The New York Times* ran an article titled "March of the Machine Makes Idle Hands," which featured experts predicting that a new invention—factory machinery that ran on electricity—would soon make manual labor obsolete. After World War II, as more factories began to install manufacturing robots, it again became conventional wisdom that workers were doomed. Marvin Minsky, the MIT researcher typically credited as the father of artificial intelligence, was reported to have said in 1970 that "in from three to eight years we will have a machine with the general intelligence of an average human being."

These fears never materialized. But today, AI anxiety is burning bright again, fueled by popular books like Martin Ford's *Rise of the Robots* and Erik Brynjolfsson and Andrew McAfee's *The Second Machine Age,* both of which made the case that AI was going to fundamentally change society and transform the global economy. Academic studies of the future of work, like an Oxford University study that estimated that as many as 47 percent of U.S. jobs were at "high risk" of automation within the next two decades, added to the sense of impending doom. By 2017, three in four American adults believed that AI and automation would destroy more jobs than they would create, and a majority expected technology to widen the gap between the rich and poor.

I spent much of 2019 reporting on these changing attitudes, being careful to keep an open mind to the possibility that these fears were exaggerated. After all, unemployment in the United States was still near a record low, and while corporate executives were chattering among themselves about AI and automation, there wasn't much obvious evidence that it was taking a toll on workers yet.

Then Covid-19 arrived. In the spring of 2020, much of the United States entered shelter-in-place lockdowns, and my phone began lighting up with calls from tech companies telling me how the pandemic was affecting their plans for automation. The difference, now, was that companies *wanted* to publicize their efforts to automate jobs. Robots don't get sick, after all, and companies that could successfully replace humans with machines could continue making goods and providing services even while the virus was raging. Consumers were excited about automation, too, because it reduced the need for human contact.

The pandemic gave companies the cover they needed to make huge, unprecedented strides in automation without risking a backlash. So they automated, and automated, and automated some more. Tyson Foods, the meat producer, brought in robotics experts to develop an automatic deboning system that could help it keep up with demand for chicken and other meats. FedEx started using package-sorting robots to fill in for sick and absent workers in its shipping facilities. Shopping centers, apartment complexes, and grocery stores splurged on cleaning and security robots to keep their stores sanitized and safe, creating shortages among those robots' suppliers.

In all, Covid-19 seemed to speed up the automation timeline by years, if not decades. McKinsey, the giant consulting firm, dubbed it "the great acceleration." Microsoft CEO Satya Nadella claimed that the company had experienced "two years' worth of digital transformation in two months." In March 2020, a survey by the accounting firm EY found that 41 percent of corporate executives were investing more in automation to prepare for a post-coronavirus world. David Autor, an MIT economist and leading automation expert, called the pandemic an "automation-forcing event," and predicted that it would usher in technological trends that would persist long after the virus was gone.

The pandemic has shown us some of the benefits of automation more clearly than any Davos panel could have. Robots and AI allowed companies to keep providing essential goods and services, even as more workers called in sick. Pharmaceutical companies used AI and automated manufacturing to accelerate their search for effective treatments and vaccines. And billions of people, stuck at home and fearful of close contact, relied on the automated, AI-powered services provided by companies like Amazon, Google, and Facebook to keep their shelves stocked and their social lives intact.

At the same time, Covid-19 has also demonstrated some of the limits of automation, and the vast numbers of important tasks we can't yet outsource to machines. We began talking about "essential workers," people whose services were necessary for society to function, and we noticed that many of those people worked not in tech or finance or some other high-prestige field, but in relatively unglamorous industries like nursing, auto repair, and agriculture. We also noticed that some activities didn't lend themselves well at all to virtualization. After a few months stuck indoors with screens as our only social conduits, many of us felt a strong pull back to the physical world. Some students stuck taking virtual classes started complaining that they weren't learning anything or having any fun. White-collar workers confined to their homes began itching to return to the office, where they could more easily collaborate with their teams and advance their careers. (One tech worker I know grumbled that "nobody is getting promoted over Zoom.") People who had been satisfied with virtual interactions during the pandemic's early months began flouting social distancing rules in order to eat at restaurants, drink at bars, and attend concerts and church services with their friends.

Machines, it turned out, could not offer an adequate substitute for human connection, or give us what we needed to get ahead. And maybe they never will.

After spending several years studying the past and present of AI and automation, I've found it hard to keep believing in the naive, utopian narrative that claims that these tools are leading us down a well-manicured path to progress and harmony. But I've also found the most dystopian, fatalistic version of the AI story—which claims that intelligent machines are destined to take over the world, and that we can't do anything about it except make peace with our own obsolescence—fairly unsatisfying.

For starters, both the optimists and the pessimists tend to talk about AI and automation in a strangely farsighted way. They focus on the effects these technologies will have years or decades in the future, and neglect to examine the effects they are *already* having.

Whether we realize it or not, most of us interact with dozens of AIs every day—the machine learning models that rank our social media feeds and power our interactions with virtual assistants like Alexa and Siri, the dynamic pricing software that determines how much we pay for hotel rooms and airline tickets, the opaque algorithms that are used to determine eligibility for government benefits, the predictive policing algorithms that law enforcement agencies use to patrol our neighborhoods. All of these systems are vitally important, but few of them get nearly as much scrutiny as the question of whether long-haul truckers will lose their jobs to self-driving eighteen-wheelers.

And while the mainstream AI and automation debate spends a lot of time talking about AI's impact on narrow measures of economic health, like productivity growth and unemployment rates, it tends to ignore more subjective questions, like whether all of this technology is actually improving people's lives. As experts like Cathy O'Neil, Safiya Umoja Noble, and Ruha Benjamin have observed, badly designed AI can harm vulnerable and marginalized groups even when it "works," by subjecting them to new forms of data-gathering and surveillance and encoding historical patterns of discrimination into automated systems. This harm can take many forms—a résumé-screening algorithm that learns to prefer men's qualifications to women's, a facial-recognition system that has a hard time correctly identifying gender nonconforming people, a predictive risk-modeling system that learns to charge higher interest rates to Black loan applicants—and any responsible discussion of AI and automation needs to grapple with these issues, too.

My biggest problem with the mainstream AI debate, though, is that both sides tend to treat technological change as a disembodied natural force that simply *happens* to us, like gravity or thermodynamics. Both the optimists and the pessimists talk about "algorithms curing diseases" or "robots taking jobs," as if machines can be programmed with both sentience and career ambition. Neither side does a good job of acknowledging that humans are waking up every day and making decisions about how to design, deploy, and measure the effectiveness of these systems.

I hear the "automation is destiny" argument all the time—especially in Silicon Valley, where people tend to talk about technological progress as a speeding train we either have to climb aboard or get run over by—and I get why it's tempting to believe. For a long time, I believed it myself. But it's wrong. And deep down, we all know it's wrong.

From the very first time a *Homo sapiens* rubbed two sticks together to make a fire, technological change has always been driven by human desires. The printing press, the steam engine, social media—these things didn't appear out of nowhere, fully intact and integrated into society. We designed them, created laws and norms around them, and decided whose interests they should serve. Innovation is not an irreversible phenomenon, either, and previous generations have successfully fought to limit the spread of harmful tools such as nuclear weapons, asbestos insulation, and lead paint, all of which represented technological progress in their day.

Whether you think AI and automation will be great or terrible for humanity, it's important to remember that none of this is predetermined. Executives, not algorithms, decide whether to replace human workers. Regulators, not robots, decide what limits to place on emerging technologies like facial recognition and targeted digital advertising. The engineers building new forms of AI have a say in how those tools are designed, and users can decide whether these tools are morally acceptable or not.

This is the truth about the AI revolution. There is no looming machine takeover, no army of malevolent robots plotting to rise up and enslave us.

It's just people, deciding what kind of society we want.

• • •

This book is not an argument that robots will take all of the jobs, some of the jobs, or none of the jobs. It's not a rant about the horrors of technological capitalism or a rumination about how we'll coexist with machine intelligence. I'm not going to predict when the singularity will arrive or tell you how to get rich building an AI start-up.

This is a book about how to be a human in a world that is increasingly arranged by and for machines. It's an attempt to persuade you that the key to living a happy, rewarding life in the age of AI and automation is not competing with machines head-on—learning to code, optimizing your life, eliminating all forms of personal inefficiency and waste—but strengthening your uniquely human skills, so you're better equipped to do the things machines *can't* do.

If you've ever felt like the world was zooming past you, or worried you have no chance of keeping up with technological change, my hope is to convince you otherwise. I want to help you keep your job. I want to help you build a healthier relationship to technology at home, and coexist peacefully with the algorithms that are trying to change what you buy, where you focus your attention, and how you view the world.

Ultimately, I want to pry our conversation about technology away from the binary poles of euphoria and terror, and foster a more honest discussion of what's coming, and what we can do about it.

Part 1, "The Machines," is an attempt to set the table. I'll draw on my interviews with experts, my reading of books and research papers, and about three centuries' worth of industrial history to explain why I believe that AI and automation are already having deep, transformative effects on our society, and why we should expect those changes to accelerate in the years ahead. I'll push back on some conventional wisdom about how machines replace workers, and explain why I fear that we've been worrying about the wrong kinds of robots.

Part 2, "The Rules," is the advice part. I'll lay out nine concrete steps you can take to prepare for the future, by protecting your own humanity and capitalizing on your most human qualities, while avoiding some of the harmful effects of today's technology. I'll show examples of people who have successfully navigated technological change this way for centuries, and explain how to apply their lessons to your own life and career.

By the end, I hope you'll share some of my concerns about AI and automation, and the economic, political, and societal challenges they could create in the coming years. But I also hope you'll feel more confident about meeting those challenges. Ultimately, my goal is to convince you that it's possible to become the type of person who has nothing to worry about: a person whose humanity makes them impossible to replace, no matter what AI can or can't do.

You will notice, as you read, that this book focuses more on the micro than the macro. There are no lengthy discussions of productivity measurement or the labor force participation rate, and I don't have a perfect set of AI policy recommendations to share. Preparing our political and economic institutions for technological change is essential, and lots of experts—including some whose work I've included in a reading list at the back of the book—have considered how we might restructure our society for the coming wave of automation. But my primary concern in this book is what *individuals*—people like you and me, with jobs and families and communities to worry about—can do.

You will also notice that I write a fair bit in the first person. That's because I'm on this journey, too. I struggle with my relationship to machines every day, and I worry constantly about my own place in an automated society. (I write for a newspaper, after all, which is not exactly the first occupation conjured by the phrase "job of the future.") Part of the inspiration for this book was selfish—I hoped I would find something, some brilliant insight or irrefutable data point, to put my own mind at ease about what the future had in store for me.

Instead, I found that the future didn't have anything in store for me, because there are no such things as "the future" or "in store." Now, as at every point in history, there are an infinite number of possible outcomes, each determined by the choices we make. If there is a robot apocalypse, it will be one of our own creation. And if this technological revolution makes the world fairer, happier, and more prosperous, it will be because we stopped endlessly theorizing and debating, took hold of our own destinies, and made ourselves futureproof.

—Kevin Roose
Oakland, California
January 2021

Part I

The Machines

One

Birth of a Suboptimist

The machine's danger to society is not from the machine itself but from what man makes of it.

— Norbert Wiener

The lights dimmed, a guitar lick boomed over the speakers, and a screen behind the stage lit up with the names of robots.

Infosec Auditor Bot—Accenture

Turbo Extractor Bot—Kraft Heinz

Web Monitor Bot—Infosys

It was April 2019, and I was in a hotel ballroom in Manhattan, watching a Silicon Valley start-up called Automation Anywhere show off its latest products to a few hundred corporate executives. These weren't the physical, *beep-boop* robots you see in sci-fi movies. They were all software bots, made of bytes and pixels, that had been programmed to take the place of human workers.

Automation Anywhere's pitch to these executives was simple: *Our bots make better office grunts than your humans.* Bots, after all, can work twenty-four hours a day, seven days a week without getting burned out. They don't take vacations, file HR complaints, or call in sick. And if you replace a human with a bot, you can, in theory, free that human up to do more valuable things.

"Twenty to forty percent of our labor workforce is trapped into acting like bridges between applications," Automation Anywhere's CEO Shukla Mihir said. When these jobs get automated, he added, "not only are people able to do higher-value work, but you are able to significantly reduce your costs."

The pitch appeared to be working. Despite its low profile, Automation Anywhere has become one of the fastest-growing start-ups in the world, with a valuation of more than $6 billion. The company's bots have been installed more than a million times, including by Fortune 500 giants like Mastercard, Unilever, and Comcast.

Several weeks earlier, I'd visited their headquarters in San Jose at Shukla's invitation. He showed me around the office, an airy single-story building with math equations stenciled on the walls, and took me to a series of four conference rooms designed as tributes to different industrial revolutions.

The first room, called "1760," was decorated as an homage to the original Industrial Revolution, with a set of factory gears hanging on the wall. The second room, known as "1840," had Edison bulbs dangling from the ceiling to commemorate the

Second Industrial Revolution of the late nineteenth century. The third room, "1969," had midcentury wallpaper and a disco light. It represented the Third Industrial Revolution—the twentieth-century wave of innovation that included technologies like the microchip, the personal computer, and the internet.

The last conference room was decorated entirely in white. It represented the Fourth Industrial Revolution—the one we're currently living through, defined by accelerating innovation in the fields of AI and automation. And the blank-slate decor, Shukla said, represented the fact that the Fourth Industrial Revolution was unfinished, and that its potential to change our lives for the better was still unfolding.

During our meeting in San Jose, Shukla told me that the age-old question about robots—will they take our jobs?—is fundamentally misguided. In fact, he believes that in many cases, robots *should* take our jobs, because our jobs are wasting our human potential.

"We're trying to pull the robot out of people, and let people achieve greater things," he said.

But in New York, onstage in front of potential clients, Shukla added a more pragmatic layer to his pitch. He told the executives that automation could cut their companies' operating expenses dramatically, and make them more profitable. He boasted that Automation Anywhere's bots weren't just single-task algorithms, but "downloadable digital workers" that could replace an entire human's workload. And he gave examples of the digital workers a company could "hire" with a few clicks: digital accounts-payable clerk, digital payroll administrator, digital tax auditor.

Then, Shukla reverted to inspirational mode, and unspooled a grandiose vision statement about the future of AI and automation, similar to the ones many of his fellow technologists have offered. It's an optimistic portrait of a win-win future, in which machine intelligence frees us from mundane labor, boosts our economy, and allows us to solve our biggest societal problems.

"I imagine a world where a hundred years from now, we'll be skiing on the slopes of Mars," he said. "Two hundred years from now, we'll be surfing on the rings of Saturn. And five hundred years from now, we'll be tapping black holes as an energy source."

Shukla paced the stage, winding up for his big finish.

"This is the potential of the human race," he continued. "But we cannot do that if forty to seventy percent of our workforce is used like robots. We must liberate human intellect!"

When I told people I was writing a book about AI and automation, I got two types of reactions.

My more tech-skeptical friends and colleagues generally approved. They'd been hearing the gloomy predictions about job-killing robots, and it worried them. They wanted me to confirm their fears of a looming automation crisis, and affirm their suspicions that even if AI didn't cause mass unemployment, it would bring new harms—creepy surveillance, runaway self-driving cars, brain-melting social media apps—that would outweigh its benefits.

Among people in Silicon Valley, though, more typical was the reaction I got from Aaron Levie, the CEO of the enterprise software company Box.

"Oh God," he said. "Please tell me you're not writing one of those 'robots are taking all the jobs' books that makes everyone terrified and depressed."

Like Mihir Shukla of Automation Anywhere, Levie believes that robots will ultimately be good for workers. He's annoyed by what he sees as alarmist media coverage of new AI technologies, and he thinks we're all worrying over nothing.

Before setting out to answer the "what do we do about AI?" question, I wanted to give this argument its due, and engage with people like Levie and Shukla in good faith. So I talked with a number of AI optimists—people who believe that these technologies will ultimately create many more positive than negative effects—and distilled what they told me into four big claims.

1. "We've been here before, and it turned out fine."

First, the optimists argue, hundreds of years' worth of evidence proves that fears about automation are generally unfounded, and that, while technology does destroy some jobs, it always creates new jobs to replace the old ones, and raises our standard of living in the process.

Sure, the Industrial Revolution cost some farmers their jobs, the optimists say. But it created millions more jobs in factories, and made entirely new categories of consumer goods cheaper and more accessible. This pattern, they say, has repeated throughout human history. Electric lights made lamplighters obsolete, but they created an entire category of new, electrified gadgets that people needed to produce, sell, and repair. Home refrigeration made ice salesmen obsolete, but it created many more jobs for grocers, restaurateurs, and farmers.

"Technology has progressed nonstop for 250 years, and in the U.S. unemployment has stayed between 5 to 10 percent for almost all that time, even when radical new technologies like steam power and electricity came on the scene," writes one such optimist, Byron Reese, a futurist and the author of *The Fourth Age*.

The people making this argument often cite today's economic data to bolster their point. Often, they refer to the "productivity paradox"—the fact that U.S. productivity growth has actually slowed over the past several decades, which is the opposite of what you'd expect to see if mass automation were making companies much more efficient, and destroying jobs left and right.

Ultimately, they say, there's no evidence that this technological shift is any different from the ones that came before it, and we should let the past reassure us about the future.

2. "AI will make our jobs better, by doing the boring parts for us."

Second, the optimists say, technology usually doesn't replace workers. Instead, it improves their jobs, by freeing them from the tyranny of repetitive, mundane tasks and allowing them to focus on more rewarding, higher-value work.

"AI Is Coming for Your Most Mind-Numbing Office Tasks," *Wired* declared in a 2020 article, noting that AI applications were being deployed inside big companies to do grunt work like data entry, document formatting, and summarizing long reports.

Optimists often cite examples of professionals who have already outsourced much of their drudgery to computers, such as doctors who use electronic medical records to do much of their routine record-keeping so they can focus on talking to patients, lawyers whose legal-research software allows them to spend more time interacting with clients, or architects whose computer-assisted design software saves them hours of pixel-pushing monotony.

These jobs aren't threatened by automation, the optimists say, because there are still plenty of things a human doctor, lawyer, or architect can do that a machine can't. And the AI that will emerge in the next few years will eliminate even more dull and repetitive work, and free us up to do the things we actually enjoy doing.

3. "Humans and AI will collaborate, not compete."

Optimists also argue that much of today's AI is designed to work *with* humans, rather than substituting for them, and that we should think of our relationship with AI as a collaborative opportunity, rather than a competitive threat.

In *The Second Machine Age,* Erik Brynjolfsson and Andrew McAfee suggest replacing the phrase "race against the machines" with "race *with* the machines." Paul R. Daugherty and H. James Wilson, two executives at the consulting firm Accenture, write in their book *Human + Machine* that human-AI collaborations will be a cornerstone of the twenty-first-century economy.

"AI systems are not wholesale replacing us," they write. "Rather, they are amplifying our skills and collaborating with us to achieve productivity gains that have previously not been possible."

One example these types of optimists often cite is Garry

Kasparov, the chess grandmaster who famously lost a series of games to IBM's Deep Blue computer program in 1997. After being beaten by Deep Blue, the legend goes, Kasparov realized that human chess players and computers would be better if they worked together. So he started promoting "freestyle chess," a type of game in which each player could consult with a computer program, and combine the machines' insights with their own expertise. These human-computer teams, Kasparov wrote, would be "overwhelming" when pitted against computers alone.

The same principle will apply, the optimists argue, to professionals in all kinds of fields. Doctors will consult with machine learning models before diagnosing diseases, judges will use recidivism algorithms to inform their sentencing decisions, and journalists will apply human touches to machine-generated first drafts. In all of those instances, the optimists say, humans and AI working together will achieve bigger and better things than either could achieve on their own.

4. "AI won't cause mass unemployment because human needs are limitless. In the future, we'll come up with new jobs we can't even imagine today."

The fourth argument the optimists generally make is that the pessimists are, essentially, failing to use their imaginations. Just a few decades ago, they say, many of the biggest companies in the world—including Facebook, Google, and Amazon—didn't exist. Until recently, there was no such thing as a YouTube creator, a search engine optimization expert, or a professional esports player.

AI, they argue, is already creating new kinds of jobs in fields like data science, precision medicine, and predictive analytics. And in the future, as AI improves, it will generate even more openings for human ingenuity. Maybe we'll all want personal trainer robots to follow us around, reminding us to eat healthier and exercise more. Maybe we'll have cities filled with connected sensors that can dynamically adjust traffic patterns to avoid congestion, or spot disease outbreaks by analyzing our wastewater. Maybe in addition to self-driving cars, we'll build self-driving restaurants, which will shuttle us from place to place as we dine. All those new projects will require humans—not just to write the code, but to give the advice, install the sensors, and provide the hospitality.

We've always been good at coming up with new, interesting work for ourselves as technology opens new doors, the optimists say, and our bottomless desires will keep us from running out of things to do.

After researching these claims, and examining the evidence AI optimists generally cite to support them, the position I landed on was neither total optimism nor total pessimism.

It's more like "suboptimism"—a word I made up to convey my belief that while our worst fears about AI and automation may not play out, there are real, urgent threats that require our attention. If I had to rank myself on a 1-to-10 worry scale, with 1 being "AI will cause no economic or societal problems whatsoever" and 10 being "AI will destroy us and everything we hold dear," I'd probably hover around a 7.

I am much less worried—maybe a 2 or a 3—when it comes to the technology itself. I still believe that well-designed AI and automation could radically improve many people's lives.

Self-driving cars and trucks alone could save hundreds of thousands of lives a year from fatal accidents, which would be a good thing even if it resulted in truckers and taxi drivers losing their jobs. Precision medicine—a new, personalized approach to disease treatment and prevention that combines AI, big data analysis, and genomics—could help us find new, lifesaving treatments for debilitating diseases. There are a million more ways AI could improve our futures, from the serious (more efficient energy consumption) to the fun (new forms of adaptive, AI-powered video games).

I am much more worried, though—maybe an 8 or a 9—about the humans who are designing and implementing all of this new technology. I've seen that AI is being eagerly embraced by profit-hungry executives and starry-eyed entrepreneurs, many of whom are deliberately underselling the risks of harm and displacement for workers. I know that many bosses are using AI to micromanage and surveil their employees and that, as a result, many jobs are getting harder and more precarious instead of easier and more rewarding. I know that flawed and biased data sets will result in flawed and biased AI, and that the overwhelming homogeneity of the engineers who build today's AI will likely result in systems that disproportionately harm marginalized groups, including women and racial minorities. I fear that AI will become increasingly useful for authoritarian governments looking to repress vulnerable populations and suppress political dissent. And I shudder to think of the privacy and human rights abuses that AI technologies like facial recognition will facilitate.

I'll admit that part of my suboptimism is a gut reaction,

informed by my years of covering the technology industry and watching it fall short of its ideals.

But it's also based on what I found out about the case for AI optimism—and why each of its main points is weaker than the optimists think.

Let's start with the first optimistic claim:

"We've been here before, and it turned out fine."

The first thing I learned is that many optimists have not done their history homework. Because while many of them claim that the Fourth Industrial Revolution will be great for human-kind, they rarely note that, for many people, the first three weren't all that great.

In the eighteenth and nineteenth centuries, as the United States and Britain industrialized, workers routinely faced bru-tal conditions in overcrowded and unsanitary factories, and were often subjected to long hours and horrendous exploita-tion. Some of the harshest conditions were faced by child la-borers, who were paid pitifully small wages, packed into squalid boardinghouses, and abused when they failed to meet their bosses' standards. The Second and Third Industrial Rev-olutions went more smoothly for workers, in part because of the labor protections that emerged out of the backlash to the original Industrial Revolution. But they still had plenty of problems. The Second Industrial Revolution created the Gilded Age, a period of American history during the late nine-teenth century that was marked by staggering corruption, bloody labor clashes, bitter racial injustice, and soaring income inequality. And the advances in communication technology during the Third Industrial Revolution generated huge productivity gains, but they also facilitated a new 24/7 work culture and introduced new sources of anxiety into white-collar workplaces, leading to unprecedented levels of burnout and job-related stress.

History suggests that while periods of technological change often improve conditions for elites and capital owners, workers don't always experience the benefits right away. After the onset of the Industrial Revolution in the 1760s, for example, Britain's gross domestic product and corporate profits soared almost immediately, but it took more than fifty years, by some estimates, for the real wages of British workers to rise. (This gap, which was described by Friedrich Engels in "The Condition of the Working Class in England," is known among economists as "Engels's Pause.") And it meant that by the time most of the workers who actually participated in the Industrial Revolution saw the fruits of their increased productivity, many of them were either retired or dead.

Some economists have suggested that we may be in another Engels's Pause today, as wages stagnate while corporate profits soar. And several recent studies have cast doubt on the argument that automation always creates more jobs than it destroys.

In particular, two economists, Daron Acemoglu of MIT and Pascual Restrepo of Boston University, have found that over the past several decades, automation has been destroying jobs faster than it has been creating them. They found that from 1947 until 1987, the optimist's take on automation was essentially correct: In industries that employed automation, job destruction and creation (what they call "displacement" and "reinstatement") happened at roughly the same rate. But from 1987 to 2017, they found, displacement in those industries dramatically outpaced reinstatement, and the new jobs that were created were generally high-skill jobs that many workers couldn't access. In other words, while displaced workers in the past could take solace in the knowledge that a new job would soon be created for them, many of the jobs being destroyed by AI and automation today likely aren't coming back.

Automation also tends to disproportionately affect people in low-income occupations, and exacerbate existing racial and gender disparities. A 2019 McKinsey report projected that Black men would be displaced by automation at a significantly higher rate than white or Asian men, in part because they are overrepresented in occupations that have high automation risk, such as truck operators, food service workers, and office clerks. (Black women would fare slightly better, the report projected, because they are overrepresented in industries like nursing and teaching, which have a lower risk of being automated.)

All of this should worry us, and it should cause us to second-guess the optimists who look to history as a source of reassurance about the effects today's AI and automation will have. (As the Oxford University economist Carl Benedikt Frey writes, "If this is 'just' another Industrial Revolution, alarm bells should be ringing.") Plenty of people suffered during the first three industrial revolutions, and plenty could suffer during this one, too.

"AI will make our jobs better, by doing the boring parts for us."

Evaluating this claim requires first defining what is meant by "better."

It's generally true that automation makes jobs less physically demanding. The most arduous blue-collar jobs of past centuries—mining, meat processing, heavy manufacturing—have largely been taken over by machines.

It's also easy to think of examples of automation eliminating dull and repetitive tasks in white-collar jobs. In my job, for example, I used to have to spend hours transcribing audio recordings of interviews I'd conducted. It was painstaking, time-consuming work, and I hated it. Today, I upload my audio files to an automated transcription service, which uses a machine-learning-powered speech-to-text engine to transcribe my audio files in seconds. Automating my transcriptions hasn't always gone smoothly. (The app tends to make some pretty amusing errors, like the time I was interviewing a confidant of Facebook CEO Mark Zuckerberg and it transcribed "Zuck's inclination" as "sexy clinician.") But it has saved me hundreds of hours over the years, and freed up time to report and write.

But despite labor-saving innovations like these, there is no evidence that today's workers are happier than workers of previous generations. Overall rates of depression and anxiety are much higher in the United States today than they were thirty years ago, and self-reported workplace stress levels have been rising steadily for decades.

This seeming paradox—that we are not getting happier at work, despite the fact that our jobs are safer and less grueling than ever—may be explained by the observation that in addition to removing hard physical labor, automation can strip away the fun, rewarding parts of jobs that workers actually enjoy.

The historian David Nye writes that in the 1930s, as the first wave of factories began to install electricity, many workers expected it to improve their daily routines. But after the lights came on, they noticed that the biggest change to their lives was that they no longer interacted with one another. The electric machines had made what had once been a dynamic, collaborative job into routine, button-pushing drudgery.

"Human contact inside the factory—the free flow of rumors, jokes, and camaraderie—became more difficult," Nye writes. "Once there had been an easy sociability during frequent interruptions, but now managers made constant improvements in the machinery and pushed up the work pace."

This kind of transformation is now happening in white-collar workplaces, as AI and automation make it possible for companies to squeeze out all the inefficiencies and downtime that once gave workers a chance to take a breath and talk to one another.

AI and automation have also created entirely new categories of boring, repetitive jobs, many of which we don't see in the West. Mary L. Gray and Siddharth Suri have written about the rise of "ghost work," a phenomenon in which human labor, carefully concealed from the end user, is deployed to make AI and automated systems function properly. Social networks like Facebook, Twitter, and YouTube rely on armies of low-paid contractors who sift through objectionable content all day, deciding which posts to leave up and which to take down. AI assistants like Alexa are helped by "data annotators," humans who listen in on recordings of users' conversations and help the system improve over time by labeling data, correcting mistakes, and training the AI to understand accents and unusual requests. In China, "data labeling" companies have sprung up to fill a need for huge numbers of workers who spend all day doing the kinds of mundane clerical work that make AI possible—for example, labeling images and tagging audio clips. These workers reportedly earn as little as 10 yuan an hour, or about $1.47.

AI optimists are broadly correct in claiming that new technology enhances our quality of life in the aggregate, and that once we acclimate to it, we rarely want to return to the old way of doing things. (Even the most hardened Luddite, I suspect, would not be thrilled about hand-washing clothes, or getting surgery without anesthesia.)

But what the optimists miss is that we don't live in the aggregate, or over the long term. We experience major economic shifts as individuals with finite careers and lifespans, and for many people, technological change hasn't always resulted in better material conditions during their lives.

I agree with the optimists, in principle, that trying to preserve outdated norms and obsolete jobs for stability's sake is a losing battle. And I'm sympathetic to the argument that we, as a society, are often too quick to mistake change for calamity.

But any honest assessment of this topic has to acknowledge that change is hard, and that many people don't seamlessly make the jump from one technological era to another. Inevitably, some people fall through the cracks. Others find their footing eventually, but never regain the stability they once had. And still others are taken advantage of by people wielding new technology, who use it to extract more from them while paying them less. Often, these revolutions end up creating lost generations—millions of people whose lives are derailed by forces outside their control, and who never make it to the promised land, or even live long enough to know what the promised land looks like.

In short, AI and automation certainly *can* improve our lives, but it's far from a given that they will.

"Humans and AI will collaborate, not compete."

I really, really wanted to side with the optimists on this one. I love the vision of humans and AIs working together, side by side, in perfect harmony. And I'd like to believe that no matter how good a machine gets at a given task, there will always be some unquantifiable X factor that human experts can bring to the table.

Unfortunately, it doesn't appear to be true.

In study after study, researchers have found that, after reaching a certain performance threshold, AI tends to outperform not only humans, but human-AI teams. One study, a 2019 preprint meta-analysis conducted by researchers at the University of Washington and Microsoft Research, looked at a number of previous studies in which decisions made by AI systems alone were assessed against "AI-assisted" decisions made by humans. In every case, they found that the AI operating on its own performed better than the AI-human team.

"Complementary performance was not observed in any of these studies," the researchers wrote, noting that "in each case, adding the human to the loop *decreased* performance."

Even the classic example of human-AI chess teams turns out to be flawed. While Gary Kasparov's hypothesis that these hybrid teams would be superior to computers alone may have been true in an earlier, less powerful era of computational chess, it doesn't appear to be true any longer. A 2014 study led by researchers at the University of Buffalo, for example, found that while human-AI teams may once have had an advantage over chess-playing AIs, "the difference does not survive to today."

In other words, in these human-AI partnerships we've heard so much about, we're often dead weight.

"AI won't cause mass unemployment because human needs are limitless. In the future, we'll come up with new jobs we can't even imagine today."

Even though it's technically unfalsifiable, I find this claim more persuasive than any of the other points AI optimists typically make. Whenever I consider the possibility that AI and automation could make humans completely useless, I think about all of the jobs that didn't yet exist when I was a kid— jobs like app developer, social media manager, podcast producer, drone cinematographer—and I find myself wondering what new, strange-sounding work we'll come up with in the decades ahead.

Industry watchers are already seeing some new jobs emerge. Accenture, the consulting firm, surveyed one thousand large corporations in 2018 and found that AI-related jobs were being created in three categories, which they called "trainers, explainers, and sustainers." These are the people who help guide and oversee machines, explain the decisions made by algorithms to other humans, and do the messy work of integrating AI into corporate IT departments. Cognizant, a rival consulting firm, recently released a list of dozens of jobs it believes will soon be created, including "personal data broker," "augmented reality journey builder," and "juvenile cybercrime rehabilitation counselor."

The big questions, of course, are whether there will be enough of these jobs to replace the jobs that are lost to automation, and whether there will be a lengthy gap between the disappearance of the old jobs and the appearance of the new ones.

These are hard questions to answer, since we don't yet know what all of the new jobs will be, or how quickly they will arrive.

But there are other questions we can start to answer, such as:

- Will the new jobs created by technology be as stable, fulfilling, and well compensated as the old jobs they replace?
- Will the new jobs be located in the same places as the old jobs?
- Will the new jobs be accessible to people of all genders, ethnic groups, and educational backgrounds, or will white men continue to have an unfair advantage?
- Will owners share the profits from automation with workers, or will they hoard them for themselves and their investors?
- Will companies lay off employees as soon as it's technically feasible to get rid of them, or will they keep them, retrain them, and put them to work in other roles?

- Will AI researchers focus on the kinds of major break-throughs that create new, job-filled industries, or will they work on incremental advances that simply allow companies to squeeze more productivity out of their workers?
- Will there be enough social and economic support for people who can't make the jump from the old jobs to the new jobs easily?
- Will companies like Google, Facebook, and Amazon use AI to empower people, connect them to trustworthy informa-tion, and improve their quality of life? Or will they use it to amplify division, spread lies and conspiracy theories, and build inescapable surveillance networks?

Notably, none of these are questions about machines. They're all about people. And the way our politicians, business leaders, and technologists answer them will determine whether AI and automation are seen as a destructive force, a humani-tarian blessing, or something in between.

Which brings me back to my suboptimism.

The good news, and the reason I'm not totally skeptical of AI's potential, is that we still have the power to determine how these technologies are developed. And if we do it right, the results could be incredible. Designed and deployed correctly, AI could help us eliminate poverty, cure disease, solve climate change, and fight systemic racism. It could move work to the periphery of our lives, and give us back time to spend with the people we love, doing the things that give us joy and meaning.

The bad news, and the reason I'm not as optimistic as many of my friends in Silicon Valley, is that many of the people leading the AI charge right now aren't pursuing those kinds of goals. They're not trying to free humans from toil and hardship; they're trying to boost their app's engagement metrics, or wring 30 percent more efficiency out of the accounting department. They are either unaware of or unconcerned with the ground-level consequences of their work, and although they might pledge to care about the responsible use of AI, they're not doing anything to slow down or consider how the tools they build could enable harm.

Trust me, I would love to be an AI optimist again. But right now, humans are getting in the way.

Two

The Myth of the Robot-Proof Job

We humans are neural nets. What we can do, machines can do.

—GEOFFREY HINTON,
computer scientist and AI pioneer

A few years ago, I got invited to dinner with a big group of executives. It was an unusually fancy spread—expensive Champagne, foie gras, beef tenderloin—and as our entrées arrived, the conversation turned, as it often does in these circles, to AI and automation.

In particular, the executives wanted to know which jobs were robot-proof. What could humans do, they asked, that machines wouldn't eventually do better?

Manufacturing was clearly out, they agreed. So was retail, clerical work, and trucking. One executive, who worked in the healthcare industry, said that AI would replace radiologists, and possibly dermatologists. Another said that many entry-level finance and consulting jobs would become obsolete. A third said that any job that was "comfortable" was at risk of being automated. (I was trying to be polite, so I didn't ask whether this guy's definition of "comfortable" included people whose jobs involved drinking Champagne and eating foie gras at a work dinner.)

When it was my turn to offer a suggestion, I froze. I suspected robot-proof jobs must exist. And I had heard plenty of experts suggest that certain professions—nursing, teaching, data science—were immune from automation. But I had also heard of start-ups that were trying to automate those very same jobs. Eventually, I came up with some platitude, about how jobs that required creativity and complex problem-solving would be hard for machines to replicate. But I knew I'd punted.

After that dinner, I started looking more deeply into the research about job automation. And I learned that the premise of that entire dinner-party conversation was flawed, because there is no such thing as an inherently robot-proof job.

Just consider some of the things we've thought were impossible for machines to do in the past.

In 1895, Lord Kelvin, a well-known British physicist, shot down the idea that airplanes would ever replace hot-air balloons as the world's aerial vehicle of choice, saying that "heavier-than-air flying machines are impossible."

Eight years later, the Wright brothers flew at Kitty Hawk, and the balloonists' days were numbered.

In 1962, Yehoshua Bar-Hillel, an Israeli mathematician and language expert, dismissed the idea that computers could be taught to translate foreign languages, writing that "there is no prospect whatsoever that the employment of electronic digital computers in the field of translation will lead to any revolutionary changes."

This one took longer to disprove, but as of 2018, Google Translate was processing 143 billion words a day, greatly reducing the demand for human interpreters.

My favorite bad machine prediction came in 1984, when *The New York Times* ran a story about the introduction of automated ticket machines at airports. The article quoted experts who were very, very skeptical that computers would ever replace human travel agents. The owner of one travel agency was quoted as saying, "What happens if you just press the wrong button?"

The owner wasn't being defensive or dense; he *literally could not imagine* any scenario in which people would ever entrust a major purchase like plane tickets to a computer. Of course, most people book plane tickets online now, and the number of humans employed as travel agents has fallen dramatically.

Keep in mind: These aren't random, uninformed bystanders lobbing bad predictions from the peanut gallery. These are leading experts in their fields, with access to better data and more insider knowledge than almost all of their contemporaries. And still they blow it, again and again and again.

In fact, when it comes to AI predictions, expertise may not be much help at all. A 2014 study by Oxford University researchers compiled six decades' worth of technologists' predictions about the timeline of AI progress, and compared them to predictions made by amateurs during the same period. The researchers concluded that there was no significant difference in accuracy between the two groups, writing that "the AI predictions . . . seem little better than random guesses."

I'm not maligning experts, and I'm not even opposed to the idea of trying to predict the trajectory of technological change. (If I were, I wouldn't be writing this book.) But I am worried that one specific type of mistake—in particular, the bias that causes us to overestimate our own abilities and underestimate the abilities of machines—could lull us into a dangerous sense of security.

For their book *The Future of the Professions,* Richard and Daniel Susskind interviewed professionals in various fields, including law, medicine, and finance, about what they thought the future held for their industry. The authors found that, although most people predicted that AI and automation would radically reshape their fields and put some of their colleagues out of work, they almost all believed their own jobs were safe.

This isn't an isolated phenomenon. A 2017 Gallup survey found that although 73 percent of U.S. adults believed that AI will "eliminate more jobs than it creates," only 23 percent worried about losing their own jobs. All over the world, in every profession, smart people seem to have simultaneously convinced themselves that (a) AI is a massively powerful technology that will be capable of performing even complex jobs with superhuman efficiency, and (b) a machine will never, ever be able to do what *they* do.

Incredibly, this kind of denialism happens even in industries where machines are already endangering jobs. In 2019, Wendy MacNaughton, a journalist and illustrator, visited truck stops in Nevada, Utah, and Idaho to ask truckers what they thought of autonomous trucks. Despite the fact that companies have spent billions of dollars developing autonomous trucking technology, and that self-driving truck prototypes are already on American highways—and may well be out of the testing stage and into production by the time you read this—almost all of the truckers dismissed the idea as ludicrous.

"Computers taking this job is a pipe dream," one trucker told MacNaughton. "Nobody can do what we do."

Part of what's confusing people about this wave of AI and automation is that the danger zone has expanded. For decades, most automation was focused on repetitive manual tasks, which were concentrated in blue-collar manufacturing jobs, and white-collar knowledge workers largely considered themselves safe. But today, many of the most promising applications of AI and machine learning are in fields like accounting, law, finance, and medicine, which involve lots of tasks like planning, prediction, and process optimization. As it turns out, these are exactly the kinds of things AI does well.

In fact, white-collar workers may actually be *more* likely to be automated out of a job than blue-collar workers. A 2019 study by the Brookings Institution, which drew on work by Stanford Ph.D. candidate Michael Webb, examined the overlap between the text of AI-related patents and the text of job descriptions from a Department of Labor database, looking for phrases that appeared in both, like "predict quality" and "generate recommendation." Of the 769 job categories included in the study, Webb and the Brookings researchers found that 740—essentially all of them—had at least some near-term risk of automation. Workers with bachelor's or graduate degrees were nearly four times as exposed to AI risk as workers with only a high school degree. Some of the most automation-prone jobs, the study found, were in highly paid occupations in major metropolitan areas like San Jose, Seattle, and Salt Lake City.

This is radically different from the way we normally think about AI and automation risk. And it should be a wake-up call to hyper-educated knowledge workers who have historically assumed that automation was someone else's problem.

Wall Street traders learned a hard lesson about their own replaceability many years ago, when high-frequency trading algorithms and computerized stock exchanges wiped out thousands of jobs for human traders on the exchange floors. Now the machines are aiming for other divisions. In 2017, JPMorgan Chase began using a software program called COIN, which uses machine learning to review certain types of financial contracts. It used to take humans more than three hundred thousand hours every year to review all of those documents. Now it happens nearly instantaneously. Many top financial firms use Kensho, an AI-based data analytics platform that automatically docs the kind of nuts-and-bolts financial analysis that used to require armies of Wharton graduates. A 2019 report by Wells Fargo estimated that as many as two hundred thousand finance employees will lose their jobs over the next decade, thanks to tools like these.

Medicine is undergoing a machine makeover, as AI learns to do much of the work that used to require trained human specialists. In 2018, a Chinese tech company built a deep learning algorithm that diagnosed brain cancer and other diseases faster and more accurately than a team of fifteen top doctors. The same year, American researchers developed an algorithm capable of identifying malignant tumors on a CT scan with an error rate twenty times lower than a human radiologist.

Lawyers aren't out of the woods, either. In a 2018 study, twenty top U.S. corporate lawyers were pitted against an algorithm developed by an AI start-up called LawGeex. Their task was to spot legal issues in five nondisclosure agreements—a staple of basic contract law—as quickly as possible. The algorithm crushed the lawyers with an average 94 percent accuracy rate, compared to the average human accuracy rate of 85 percent. Starker still was the difference in billable hours: it took the lawyers an average of ninety-two minutes to complete the challenge, whereas LawGeex's AI finished in twenty-six seconds.

Even computer programmers, long seen as the white-collar workers with the best job opportunities, are at risk of automation. "No-code" and "low-code" development interfaces, which allow non-programmers to create applications, as well as centralized service providers like Amazon Web Services, have made it possible for companies to write software and maintain technical infrastructure with fewer humans than ever before. Even AI engineers may be automating themselves out of jobs. In 2017, Google released AutoML, a suite of tools that uses machine learning models to build and train other machine learning models. The results of Google's initial tests were impressive: after being instructed to build a neural network capable of carrying out a common image-labeling task, Google's AI was able to build and train a model that was more accurate than the one Google's own engineers had programmed.

And journalists? Forget it. Many of us are eminently automatable, especially those of us whose output tends to be more routine and predictable. In 2020, several publications began experimenting with GPT-3, an advanced AI program developed by the nonprofit research lab OpenAI. The program, which takes a prompt and uses machine learning to complete it, was able to produce long, cogent pieces of writing that amazed human editors with their clarity and style. One publication, *The Guardian,* used GPT-3 to write an entire op-ed about the future of AI and machine learning, and concluded that "overall, it took less time to edit than many human op-eds."

This isn't to say that machines will replace all white-collar workers, or even most of them. But it's a warning that elite college degrees, impressive LinkedIn profiles, and six-figure salaries are no longer shields against obsolescence.

Two other types of jobs that are often suggested to be impossible to automate are "compassion" jobs and "creative" jobs—jobs that involve caring for people and coming up with new ideas.

But researchers and entrepreneurs are successfully automating some limited kinds of work in both of these domains. Stanford researchers recently developed Woebot, a "chatbot therapist" that uses machine learning and standard cognitive behavioral therapy to talk users through their problems—an approach that peer-reviewed studies showed significantly reduced symptoms of depression and anxiety in users. In Japan, "carebots" are being developed to help older people remember to take their medicines, help them move and feed themselves, and provide them with a sense of companionship. These robots aren't capable of fully human interactions, but they might not need to be. Early research on the effectiveness of elder-care robots, including a 2019 study led by researchers at the

University of Auckland, New Zealand, found that they can be just as effective as humans at interacting with people experiencing dementia.

In addition, some of the skills we thought were unique to humans—such as the ability to read and interpret emotions—may actually be replicable by machines. In fact, there is a whole subdiscipline of computer science, called "affective computing," that uses AI to analyze speech and facial microexpressions in order to determine people's emotional states. And while the effectiveness and accuracy of these systems is being hotly contested, some of them have done impressively well. A 2019 study led by Eva G. Krumhuber at University College London found that an AI classifier was better than humans at correctly identifying emotions in a series of posed videos, and roughly as good as humans as correctly identifying spontaneous, non-posed displays of emotion.

As for creative jobs, it may be a while before AIs are kicking Leonardo da Vinci out of the Louvre. But early experiments in computer-assisted art making have shown some promise. I recently went to a gallery show in which every painting was generated by AI using a machine learning technique known as a "generative adversarial network." The pieces were haunting and eerie and beautiful, and the collectors in the room snapped them up, some paying thousands of dollars per piece.

AI is making strides in other creative fields as well. Algorithms now write screenplays, design video game levels, and generate architectural blueprints on their own, and studies have shown that humans frequently prefer the machine-generated creative output to works by experienced humans.

The journalist Clive Thompson recently wrote about Juke-deck, an AI-based music-writing tool that allows users to generate new compositions on the fly. Thompson wrote that although Jukedeck might not replace headline acts, it could cut deeply into the ranks of studio musicians who create soundtracks and stock music libraries.

"The tune wasn't brilliant or memorable, but it easily matched the quality of human work you'd hear in videos and ads," Thompson wrote of the demo track Jukedeck produced for him. "It would take a human composer at least an hour to create such a piece—Jukedeck did it in less than a minute."

There is one more fundamental problem with the "robot-proof job" discussion—namely, it places too much emphasis on what our job title is, and too little emphasis on what qualities we bring to that job.

Most of the existing studies of AI and automation have analyzed the automation risk of broad occupational categories, giving all teachers, all architects, and all factory workers the same chance of getting wiped out. There is even a website, WillRobotsTakeMyJob.com, where you can plug in your occupation and see your supposed risk of automation-linked job loss. (I got 11 percent for "reporters and correspondents," which, frankly, seems optimistic.)

In reality, most jobs can be done in ways that either make them very easy or very hard to automate. An "artist" could be a person who teaches art therapy classes to people with autism or a guy who draws goofy caricatures at Six Flags. A "doctor" could be a beloved small-town pediatrician or a diagnostic radiologist who spends all of his or her time analyzing scans in a lab. A "journalist" could be an investigative reporter who exposes malfeasance and wrongdoing at the highest levels of government or a person who summarizes corporate earnings reports for a newswire. Despite having the same title, these people do not have the same risk of AI replacement.

Another problem with these occupation-level studies is that jobs that seem routine and predictable often aren't.

For example, consider airport TSA agents. Every day, they tell passengers to take the liquids and laptops out of their bags, guide them through the body-scanning machines, and check their luggage for prohibited items. Unskilled, routine job, right? Easy to automate? Except it turns out that TSA agents do more than just stare at the X-ray machine all day. They manage unexpected situations and anomalies, like passengers with medical issues who can't go through the body scanner, or people traveling without their IDs. They track down lost items, put nervous travelers at ease, and look for subtle signs of passenger behavior that might indicate a security threat. They do a million other tiny tasks that don't appear anywhere in their job description, but without which any airport would grind to a halt. It will likely be harder for machines to replace TSA agents than the data would indicate.

Some jobs are also *more* robotic than they first appear. Take fashion design, for example. Designing clothes might appear to be a purely creative job that would be impossible for computers to perform. But a lot of modern fashion design— especially at "fast-fashion" chains and e-commerce brands— consists largely of pattern recognition and data analysis, and figuring out how to create variations on an item that is already selling well. As it turns out, this is the kind of task that AI can do very well. In fact, several companies are already using AI to generate fashion designs. In 2017, an Amazon research team developed a machine learning algorithm that analyzes images of garments in a particular style and learns to generate new garments in that style. Glitch, an AI fashion company started by two MIT graduates, sells pieces that are entirely designed by deep learning algorithms.

Will AI spare all TSA agents, or replace all fashion designers? Of course not. But the fallout from automation probably won't be as tidy as watching some occupations go extinct while others survive without a scratch.

In short, what I should have told the executives at the fancy dinner was that they were asking the wrong question. Robot-proof jobs don't exist, and our job titles are not our destiny.

When it comes to avoiding machine replacement, what we do is much less important than *how* we do it.

Three

How Machines Really Replace Us

Some technologies come in disguise . . . they do not look
like technologies, and because of that they do their work,
for good or ill, without much criticism or even awareness.
—Neil Postman

There's a classic episode of *The Jetsons,* the 1960s cartoon about a family living in a robot-filled future, that illustrates the way we typically think about people being replaced by machines. In the episode, George Jetson goes off to work at his factory job. When he arrives at the factory, his boss calls him into his office and tells him that a robot named Uniblab has been brought in to do his job. (As a consolation prize, the boss offers George a job as Uniblab's assistant.)

Half a century later, this is still the stereotypical image of how workplace automation happens. You arrive at the office one day, and a robot is sitting in your seat. Your boss awkwardly tugs his collar and breaks the bad news.

This kind of one-for-one substitution still happens occasionally, like in 2019, when Walmart brought in a fleet of floor-cleaning robots and simultaneously laid off hundreds of human janitors. (According to *The Washington Post*, workers at a Walmart store in Marietta, Georgia, named their floor-cleaning robot Freddy, after the beloved human janitor it had replaced.) But Jetsons-style firings are rare, and getting rarer, for reasons having more to do with capitalist efficiency than anything else—basically, if an off-the-shelf hardware robot could replace you, it probably already has.

These days, more typical are stories like the one I heard from Jamie Lerman, an insurance salesman from New Jersey who works at a small, family-owned branch of a large national insurer. When Lerman started selling insurance a decade ago, his branch was full of agents who spent all day placing sales calls, calculating quotes for new policies, and taking care of customer billing. But new technology automated many of those functions. Today, the firm's staff is half the size it was when Lerman started, and many of the desks in the office sit empty.

"It's not that people are getting fired," he told me. "It's that when they leave, there's less and less urgency to replace them immediately. We just don't need that many people anymore."

There are even more subtle forms of automation-related job loss. Consider the following (purely hypothetical) scenarios:

1. An aerospace giant with eighty thousand employees and factories all over the country has seen its sales of new planes fall dramatically over the past several years. One of the reasons is that a twenty-person start-up in San

Francisco built an app that uses machine learning to extend the useful life of airplanes by applying predictive algorithms to figure out when certain parts need to be replaced and maintained. Armed with this software, airlines start replacing their jets less frequently, and the aerospace giant's sales miss estimates several quarters in a row. Under pressure from shareholders and its board, the company decides to shut down several plants and lay off 25 percent of its workers.

2. A trucking company has carried freight for the same big-box retailer for decades. But one day, the retailer's logistics division starts using a new AI-powered "load optimizer" to make its delivery routes more efficient and reduce the number of trucks it needs to carry the same amount of freight. The next year, the trucking company sees its delivery orders shrink by 30 percent, and it's forced to lay off a number of drivers and dispatchers.

3. A prestigious New York law firm has hired fifty law school graduates every summer for the last twenty years. The firm's biggest client, a Wall Street investment bank, has just installed an AI-based tool that automatically reviews certain types of documents and flags legal compliance issues. The program can be operated by junior bank employees making $40 an hour, rather than $400-an-hour law firm associates, so the bank dramatically scales back its use of outside counsel. The partners of the law firm, who had not factored this change into their revenue forecasts, decide to hire only twenty-five law school graduates next summer.

All of these scenarios involve job loss related to automation, even though none of them involves a direct, one-for-one substitution. If one of them happened to you, it might never even occur to you that technology was involved. All you'd see are the secondary effects: budget cuts, empty trailers, fewer job offers.

This dynamic is part of what the technology writer Brian Merchant calls the "invisible automation" problem. Merchant writes that "automation does not appear to immediately and directly send workers packing en masse." Instead, he says, its effects often appear gradually, in the form of pay cuts, unfilled openings, and higher turnover.

In fact, there are several common ways machines facilitate the replacement of human workers, none of which involves the Jetsons-style scenario.

Small Companies for Big Companies

The first is that automation allows small companies to do similar tasks as their bigger, more established rivals with many fewer human workers.

In their book *Competing in the Age of AI,* Harvard Business School professors Marco Iansiti and Karim R. Lakhani illustrate this concept using the example of Ant Group, a financial services start-up affiliated with the Chinese e-commerce giant Alibaba. Ant Group, which began as a payments platform called Alipay, is one of the most valuable private companies in the world. And it got that way, in large part, because it figured out how to replace many of the labor-intensive services offered by traditional banks with machine processes.

For example, one Ant Group affiliate, MYbank, is a lending app whose signature procedure is known as "3-1-0" be-cause of what it requires: three minutes to apply for a loan, one second for an algorithm to approve it, and zero humans. The bank has lent out hundreds of billions of dollars this way, and thanks to the consumer data it collects from Alibaba and other partners, it has kept its default rate down around 1 per-cent, well below that of many traditional lenders.

MYbank, which had only about three hundred employees as of 2018, will never have to lay off thousands of human loan officers to make way for algorithms, because it never employed thousands of human loan officers in the first place. But those jobs exist at other Chinese banks and lending firms. And it's a good bet that as MYbank continues to grow, many of those other companies will be forced to cut their payrolls to keep up.

New Behaviors for Old Behaviors

Machines also substitute for human workers by changing the way we perform certain tasks.

Take Kodak, the onetime photography giant. In 1988, Kodak was a thriving business with a stunning 145,000 workers on its payroll, including a big chunk of the population in the company's hometown of Rochester, New York. Back then, if you'd asked a Kodak executive what the biggest threats to those workers were, they probably would have said something about outsourcing or foreign competition. A really future-minded ex-ecutive might have predicted the rise of digital cameras.

But overseas competition and digital cameras didn't kill Kodak. Smartphones and social media did. As millions of peo-ple began carrying iPhones and Android devices with high-resolution cameras in their pockets, they stopped thinking about photography as a paid service that required dedicated equipment and started thinking of it as a DIY hobby. Tech companies didn't set out to kill Kodak, but by changing the underlying consumer behavior from one in which people printed out their photos to one in which they uploaded them to websites, they effectively sealed its fate. Kodak declared bankruptcy in 2012; today, it has only about five thousand employees.

It sounds strange to say that the other 140,000 jobs at Kodak were automated away, because that automation *didn't happen at Kodak*. It happened at MySpace, Facebook, Instagram, Twitter, and the other companies that provided photo-sharing tools. But when those companies adopted technology to allow users to share their photos online without ever setting eyes on a film canister, the result was people in Rochester losing their jobs.

Freelancers for Full-time

Machines also allow companies to substitute part-time, temporary, and contingent workers for full-time employees, by breaking jobs down into standardized tasks that can be performed by relative amateurs and allowing small numbers of managers to supervise large, flexible workforces.

The typical examples of this phenomenon are gig economy companies like Uber, Lyft, and Airbnb, all of which have made it possible for people with cars and spare bedrooms to compete with professional drivers and hoteliers. But a better example may be what's happened in my industry. Several decades ago, human journalists were employed at newspapers, magazines, and TV stations, and given the job of separating fact from fiction, deciding which stories were appropriate for an audience, and ranking the day's news in order of importance. They were called "editors," "producers," and "reporters," and there were tens of thousands of them, most earning a decent middle-class living.

Today, a huge number of those jobs have disappeared, and in their place sits the automation-age job title of "content moderator." Like the editors and producers of old, content moderators spend their days making sure that information being broadcast to the masses through Facebook, YouTube, Twitter, and other platforms is suitable for public consumption. They usually aren't employed by the platforms themselves, but rather contracted out through temp agencies and consulting firms. Few make much more than minimum wage. And while they sift through objectionable content all day, they have little of the training that once gave editors and producers the ability to make real-time, subjective judgments about which stories to run and which stories to kill. Instead, they follow depersonalized "content guidelines" and decision trees handed to them by their managers. The ultimate goal of the tech companies is to automate this process and replace all the human moderators with AI that can accurately detect hate speech, graphic violence, and other prohibited types of content. But in the meantime, they will just substitute low-paid contingent workers for salaried professionals.

The subtle, indirect ways automation transforms our lives and workplaces often make it hard to pinpoint the nature of any single threat. But in hindsight, we often realize that the technology that looked innocent and helpful when we first encountered it ended up having a more destructive effect.

In 1984, when TurboTax arrived, it didn't look like a job-killing robot—it looked like a piece of software that let computer geeks fill out their taxes on a PC—but it eventually forced legions of tax preparers to find new work.

In 1985, when Microsoft Excel was released, it didn't look like a job-killing robot—it looked like a spreadsheet program—but it eventually obviated the need for entire departments full of manual data entry clerks.

In 2006, when Facebook added a feature called the "news feed," it didn't look like a job-killing robot—it looked like a way to find out which of your college crushes were newly single—but it morphed into a product that distributed information to billions of people, dominating the online ad market and reducing the demand for newspapers and magazines.

It's almost certain that some of the technologies in our lives today will end up costing humans their jobs, just as these tools did. An easy lesson to draw from history is that machines disrupt our lives in ways we don't see coming. We worry about Skynet, not spreadsheets. And when the change arrives, we're often caught by surprise.

Four

The Algorithmic Manager

I felt so stifled, my brain wasn't needed anymore. You just sit there like a dummy and stare at the damn thing. I'm used to being in control, doing my own planning. Now I feel like someone else has made all the decisions for me. I feel downgraded.

—worker at a recently automated
General Electric plant in 1970

Every weekday, Conor Sprouls goes to work as a customer service representative for MetLife at a call center in Warwick, Rhode Island. When he gets to his desk, he boots up his computer, and a small blue window appears in the bottom-right corner of his screen.

This blue window is Cogito, an app-based "AI coach" that MetLife uses to keep tabs on its human customer service representatives. Every time Sprouls gets a call, Cogito listens in on the line, and provides him with real-time feedback. If the app detects that he's talking too fast, the window flashes a picture of a speedometer to remind him to slow down. If he sounds sleepy, it flashes a coffee-cup icon to perk him up. And if, for some reason, Cogito believes Sprouls isn't connecting with the customer, it might show him a heart icon—an "empathy cue"—to encourage him to mirror the customer's emotional state.

When we imagine classic workplace automation, we mostly imagine machines doing low-level grunt work overseen by humans. But in many of today's workplaces, AI has been promoted to middle management. In industries from customer service to banking and food service, software now does the supervisory work of training workers, monitoring quality, and reviewing performance—all tasks that used to fall to humans.

The idea of algorithmic bosses isn't new. In the twentieth century, "process optimization" tools were used to squeeze extra efficiency out of manufacturing workers, and workers in the service industry have decades of experience with "dynamic scheduling" software like Kronos, which is used to set workers' shifts based on a business's predicted staffing needs. But AI and machine learning have made it possible to outsource even higher-level management tasks to machines. Amazon uses complex algorithms to track the productivity of its warehouse workers, and can reportedly even automatically generate the paperwork required to fire underperformers. IBM has used Watson, its AI platform, in employee performance reviews— meaning that your bonus might be determined not simply by how you did last year, but by how the algorithm predicts you'll do next year. On-demand platforms like Uber and Lyft have dispensed with the idea of human supervision altogether, putting decisions like pay, dispatching, and dispute resolution in the hands of algorithms.

Algorithmic management has become a lucrative industry. In addition to Cogito, there are also retail-oriented AI companies like Percolata, a Silicon Valley start-up that counts Uniqlo and 7-Eleven among its clients, which uses in-store sensors to calculate a "true productivity" score for each worker. Another AI start-up, Beqom, automates the process of calculating worker pay and year-end bonuses. And Nexus AI, a "workforce management" system, allows managers to sort workers into teams based on calculated attributes like "high performers" and "good chemistry."

When I visited MetLife's call center, I was struck by the authority Cogito's software had been given, even though it was still fairly new. The app tracks the number of notifications each agent receives, which are compiled into a score and used by managers to track agent performance over time. (Agents aren't allowed to minimize their Cogito windows on their screens—if they do, the program alerts their supervisor.) Every agent still has a human manager, and although the company assured me that Cogito scores are never used as a determining factor in things like pay and performance reviews, Chris Smith, MetLife's global head of operations, told me that Cogito had allowed the company to correct underperforming employees.

"There was one associate, we were seeing her calls were going several minutes longer than the traditional call," Smith said. "By listening to Cogito, they could tell she was repeating information that didn't need to be repeated."

MetLife, which uses Cogito's software with more than 1,500 of its call center employees, says using the app has increased its customer satisfaction scores by 13 percent. And the MetLife agents I spoke to during my visit didn't seem miserable about it. (Granted, I was being accompanied by an official from the company's corporate communications team, so employees may have been on their best behavior.) Mostly, the AI seemed to be a slight annoyance, but a tolerable one.

"When it was first rolling out, there were some concerns like, 'Oh, this thing's going to be yelling at me on every call,' but we don't feel that way," Sprouls told me. "I think at the end of the day, it's just a cool piece of tech."

Another MetLife agent, Thomas, was less enamored with Cogito. "At the beginning," he said, the software "was giving me a lot of notifications, because it was getting used to my voice." In particular, the app frequently flagged him for "continuous speaking," and prompted him to allow the customer to get a word in edgewise. But even after he did, the app continued to alert him for things he didn't think he'd done wrong, like talking too fast or failing to show empathy.

"Sometimes it pops up and I just leave it, because I know I'm doing it right," he said.

Defenders of algorithmic management often point out, correctly, that many human bosses have flaws of their own. They make rash decisions. They violate boundaries and play favorites. They can be egomaniacal and cruel. In theory, automation could replace the worst human bosses, and make the good ones better by equipping them with better tools and information.

Some start-ups are working on this. Humu, a company founded by the former Google executive Laszlo Bock, is trying to use AI to improve managers. The company, whose clients include Sweetgreen and OfferUp, sends email and text message "nudges" to managers throughout the day, reminding them to do things like explain their decisions more clearly and give more direct feedback to their employees. Other apps, like Coach Amanda, Butterfly, and QStream, are building similar types of automated manager training systems.

The long-term effectiveness of these programs is still unknown. But we do know that when opaque managerial algorithms are implemented without enough human oversight, trouble often follows. Workers at Instacart and other on-demand delivery services have gone on coordinated "strikes" to protest controversial policies, like a setting in Instacart's app that applied customer tips toward mandatory minimum payments, rather than putting them on top of the minimums. Full-time YouTube creators—perhaps the most directly machine-managed humans on earth—have made a sport of dissecting the platform's all-important recommendations algorithm and complaining about its effects on their channels. A 2019 study of Uber drivers found that many of them felt frustrated and dehumanized by their algorithmic management structure, in which everything from their wages to their performance ratings was determined by opaque, inscrutable machines. Many drivers, the study found, had resorted to various forms of insubordination, such as coordinating with other drivers to game the system by artificially triggering surge pricing in a given area.

Maneuvering like this may become much more common as machines take on more power within organizations. In the future, we may all feel much more like YouTubers and Uber drivers—subjects at the mercy of a capricious machine that has the power to make or ruin our careers. Workplace AI will not only hire and fire us, but guide our daily performance, correct us when we slip up, and praise us when we do good work. "Playing office politics" will come to mean "reverse-engineering a piece of workforce management software." A "hostile work environment" might be the result of a poorly trained machine learning model rather than an abusive boss. And it will be an open question whether the best path forward for workers is to accept these machines' authority, or to challenge it.

Five

Beware of Boring Bots

Your actions indicate you intentionally misled and/or concealed information to obtain benefits you were not entitled to receive. . . . You are disqualified for benefits under MFS Act, Sec 62(b).

—letter received by a Michigan resident whose benefits were wrongfully terminated by MiDAS (Michigan Integrated Data Automated System), a fraud-detection algorithm used by the state's unemployment insurance agency

Picture the scariest robot you can imagine. Maybe it looks like the Terminator—an armed, humanoid killing machine. Maybe it looks like a self-driving car that goes rogue and starts driving people into walls and over cliffs. Or maybe it looks like one of those (admittedly terrifying) four-legged robot dogs that are featured in viral internet videos, playing soccer and doing parkour moves.

These kinds of robots are the most visible physical representations of automation in our society, and they're the kind we generally picture when we worry about being overtaken or threatened by machines.

But I believe that the bigger risks, at least in the near term, will come from forms of automation that nobody is paying much attention to.

I call these "boring bots." And I think there are two main types to watch out for.

The first are what I call "bureaucratic bots." These are the faceless, anonymous algorithms that are used by government agencies, financial institutions, healthcare systems, criminal courts, and parole boards to make important, life-altering decisions, but that rarely attract the same kind of attention or scrutiny as consumer automation products made by companies like Amazon and Google.

In her book *Automating Inequality*, Virginia Eubanks, a professor of political science at the University of Albany, examines the rise of bureaucratic bots, and the way they are being used by state and local governments to automate the process of determining low-income residents' eligibility for housing vouchers, medical assistance, and other critical benefits. These systems, she writes, are often sloppily designed and administered, and can create Kafkaesque nightmares for people trying to figure out why an algorithm kicked them off Medicaid or took away their food stamps.

Sometimes, as in the Michigan unemployment insurance case, a bureaucratic bot's mistakes are caught and reversed by humans later on. (A class-action lawsuit representing approximately forty thousand Michigan residents who were wrongfully denied benefits by the MiDAS algorithm is still pending in court; a state review later found that MiDAS had a 93 percent error rate.) But these mistakes often have life-changing consequences. In 2007, a glitch in an automated system used by the California Department of Health mistakenly cut thousands of low-income seniors and people with disabilities from their benefits. In Ohio, a yearslong project to overhaul the state's benefits processing software has resulted in thousands of residents being wrongfully denied SNAP benefits, or having necessary forms mailed to the wrong addresses. In Idaho, a flawed automated process in the state's Medicaid administration system resulted in thousands of people with mental and physical disabilities getting their benefits cut dramatically with no explanation.

Eubanks writes that putting people at the mercy of bureaucratic algorithms, with little human oversight, "shatters the social safety net, criminalizes the poor, intensifies discrimination, and compromises our deepest national values." She's right, and as critical systems like these become more highly automated, the risk of life-altering mistakes due to bad programming or inadequate supervision will only grow.

The second category of worrisome automation consists of what I call "back-office bots." Back-office bots are the software programs that can do the kinds of menial, unsexy tasks that are necessary for any large organization to function. If you work in a big company, you can probably think of someone with a generic-sounding title like operations coordinator or benefits administrator—these are exactly the kinds of people back-office bots are designed to replace.

Many of these apps fall into a category known as "robotic process automation," or RPA. Automation Anywhere, the company whose conference was detailed in this book's introduction, is a major RPA vendor, but there are others you've probably never heard of, with names like UiPath, Blue Prism, and Kryon. Collectively, these companies are worth billions of dollars, and they've been growing so quickly that even large tech companies have stepped into the RPA business. In 2019, Microsoft announced that as part of its entry into the RPA market (and fair warning, you may need a double shot of espresso to make it to the end of this sentence without falling asleep), the company was adding "an end-to-end automation solution" to its cloud-based Power Automate platform that would "couple the capabilities of UI flows with Power Automate's prebuilt connectors for more than 275 widely used apps and services that support API automation."

This is not the kind of sexy, exciting work that wins prizes at AI conferences or gets published in peer-reviewed journals. Nobody will ever make a Super Bowl commercial about a relational database plug-in. In fact, some computer scientists don't think RPA qualifies as AI at all, since it usually involves static, rule-based programs and not adaptive algorithms that learn on their own.

Still, these boring robots represent the kind of cost-saving technology that businesses are willing to pay big money for. By some estimates, RPA is one of the fastest-growing parts of the AI industry, and is expected to be a $6 billion industry by 2025. RPA providers' websites are filled with glowing "success stories" from large corporations that have used their products:

"Sprint Automates 50 Business Processes in Just Six Months"

"Dai-ichi Life Insurance Saves 132,000 Hours Annually"

"600% Productivity Gain for Credit Reporting Giant with RPA"

These case studies are carefully worded to avoid any mentions of job cuts or layoffs. (Notice that they say things like "saves 132,000 hours" rather than "replaces 65 people in the finance department.") But people who pay attention to the RPA industry—and there are only a handful of them because, and I cannot emphasize this enough, it is *incredibly* boring—have told me that job losses are almost always part of the equation.

Craig Le Clair, an analyst at Forrester Research, started hearing about RPA back in 2015. He knew companies were investing heavily in technology, but when he began talking to executives at major Fortune 500 firms, he was shocked at how much they were spending on job-replacing robots from companies no one had ever heard of.

"I was seeing enterprises spending $20 million to do this kind of automation," he told me recently. "And if you go talk to your neighbor, or anybody on the street, and say, 'Do you know what RPA is?,' the answer is no. They just have no idea."

Le Clair noticed that these RPA companies weren't doing anything particularly fancy. Most of the time, he said, they were "just building a script that does what Harry was doing in the back office." But executives loved these bots, because they plugged into their existing software programs and allowed them to automate jobs without rebuilding their entire technical infrastructure, which could take years and cost billions of dollars.

"You go to these conferences, you talk to the CFOs in a corner, and you ask them, 'So, what are you really doing with this stuff?' And they're taking people out," he said. "You can build a bot that costs ten thousand dollars a year and take out two to four humans."

Le Clair suspects that there are a lot more back-office Harrys facing unemployment at the hands of RPA than these executives are letting on—maybe millions more. And he doesn't buy the typical claim that these bots improve workers' jobs rather than eliminating them. He has noticed instances of executives publicly proclaiming that they would shift workers to other departments once their jobs were automated, then quietly laying them off weeks or months later. He and his colleagues crunched some numbers and estimated that, by the year 2030, RPA and other forms of automation could eliminate more than 20 million American jobs.

Boring bots have obvious risks for workers: job cuts, lost benefits, denied insurance claims. But they are also dangerous on a macroeconomic level—not because they are too powerful but because, in some sense, they aren't powerful enough.

Part of the reason we never experienced mass unemployment after big technological shifts of past centuries is that while transformative new technologies destroyed some jobs, they boosted productivity and created more labor demand elsewhere in the economy. Shipping containers put some long-shoremen out of work, but they made it much cheaper to transport cargo across the world, which boosted global trade and lowered prices for all kinds of consumer goods. Those low prices, in turn, enticed consumers to buy much more stuff, and created more jobs at the companies that produced all that stuff.

But in recent years, much of the automation we've gotten hasn't made us much more efficient.

In a 2019 paper, MIT's Daron Acemoglu and Boston University's Pascual Restrepo coined the term "so-so technologies" to describe the type of machine that is good enough to replace human workers but isn't good enough to generate new jobs. So-so automation, they wrote, is the kind we should really fear, because it allows employers to substitute machines for humans without the major productivity gains that could create new jobs somewhere else.

"It is not the 'brilliant' automation technologies that threaten employment and wages, but 'so-so technologies' that generate small productivity improvements," Acemoglu and Restrepo wrote.

One example of so-so automation is the grocery store self-checkout machine. As any shopper can tell you, these machines are *very* so-so. They break frequently, they often scan and weigh items incorrectly, and cashiers constantly have to be called over to key in a manual override. These machines don't make grocery stores ten times more productive, or radically increase how many groceries we buy. All they do is transfer labor from employees to customers, which allows grocery store owners to staff slightly fewer people on a shift.

Another example of so-so automation is an automated call center. Replacing human customer service agents with an automated system doesn't radically increase the company's sales, or improve the quality of its products. All it does is allow the company to shrink a "cost center"—to do the same amount of work with slightly fewer humans—and transfer the burden of problem solving to customers.

The surge in so-so automation may explain why America's economic productivity hasn't risen substantially in recent years, despite the advances being made in automation and robotics. And it means that, counterintuitively, if people losing jobs to robots is your primary concern, you should probably want robots to be more capable, not less.

Given the trouble that boring bots can create, both for the systems and programs that people depend on for basic services and for the overall labor market, it's time for us to update our mental image of AI danger. For now, as strange as it sounds, we may want to stop worrying about killer droids and kamikaze drones, and start worrying about the mundane, mediocre apps and services that allow companies to process payroll 20 percent more efficiently, or determine benefits eligibility with fewer human caseworkers.

I believe, as experts like Eubanks and Le Clair do, that we underestimate boring bots at our peril.

Part II

The Rules

Rule 1

Be Surprising, Social, and Scarce

You cannot endow even the best machine with initiative;
the jolliest steam-roller will not plant flowers.

—WALTER LIPPMANN

On June 23, 1821, a twenty-one-year-old Englishman named William Lovett arrived in London with thirty shillings in his pocket, looking for a fresh start.

Lovett was a working-class kid from Newlyn, a fishing village on the southwest tip of England. As a teen, he had apprenticed for a local rope maker, and planned to go into the rope making trade as an adult. Making rope wasn't the most prestigious job in the world, but it was steady work, and Lovett found satisfaction and purpose in it.

Unfortunately for him, the Industrial Revolution was well under way, and a newly developed technology—the metal chain—was disrupting the rope business. Rope sales were plummeting as customers opted for the stronger, more durable material that could be mass-produced by big, steam-powered machines in urban factories. After failing to find steady work as a rope maker, Lovett realized that the skills he had spent his teenage years honing were becoming obsolete.

He wasn't alone. All over England, workers were coming to terms with their own irrelevance. Industrial machinery had upended the careers of blacksmiths, farmhands, and other manual laborers, and cost tens of thousands of artisans their jobs. Some workers fought these changes head-on, including a coalition of machine-smashing textile workers in Manchester who became known as the Luddites. Others, including Lovett, started looking for a new line of work instead.

Changing careers wasn't easy, and Lovett had several false starts. He joined the crew of a fishing boat, but quickly discovered that it made him seasick. He worked for a carpenter, but was forced out of that job in short order after several of the carpenter's young apprentices complained that Lovett was unfit for the trade.

Growing more desperate, Lovett packed his bags, said goodbye to his family, and set off for London. There, he hoped, he'd discover a new future waiting for him.

Maybe you've had a moment like that—an instant when you felt the future rush past you, and worried that the skills you'd spent your life accumulating were suddenly worthless.

My William Lovett moment came in 2012. I was in my mid-twenties, covering Wall Street and the stock market for the *Times*. At the time, the newspaper industry was in steep decline, and it seemed as if jobs like mine might vanish at any moment. Lots of my friends in journalism had been laid off, lots of print publications had closed or gone online-only, and there were constant rumors about the next dominoes to fall.

One day, I read a story about a start-up that was developing an AI reporting tool based on a process called "natural language generation," or NLG. This program could take structured data—statistics from a corporate earnings report, for example, or a database of real estate listings—and turn it into a full-fledged news story in milliseconds, with no human reporters or editors needed.

These robot journalists weren't winning any Pulitzer Prizes, but they never missed their deadlines, and they were shockingly productive. The makers of one NLG app, Wordsmith, claimed that it had churned out 300 million news stories in a single year—more than every journalist on earth combined. Another app, made by a company called Narrative Science, was being used by sports websites like Big Ten Network to turn box scores and player information into auto-generated game recaps. And major media companies like the Associated Press, Forbes, and Reuters were signing up to incorporate AI reporters into their newsrooms.

When I first heard about these AI reporting apps, I dismissed the idea that they could pose a threat to human journalists. Computers, I thought, might take over the more mundane, routine tasks involved in journalism—the fact gathering, the number crunching, and the formulaic stories we all hated to write. But they would never be capable of doing the more creative, human parts of our jobs: generating story ideas, gathering quotes from reluctant sources, or explaining complex ideas in an accessible way.

But as I thought more about it, I started to worry that I'd been deluding myself. After all, I'd written plenty of formulaic news stories, including writing up corporate earnings reports and compiling new economic data. Some of my assignments were creative and complex, but for others, the goal was simply to transmit information as quickly and accurately as possible.

As I thought more about these apps, and compared them to my own output, I began to wonder whether I'd been too confident. Some days, I thought, maybe a robot *could* replace me.

For years, the conventional wisdom has been that if machines were the future, we needed to become more like machines ourselves.

When I graduated from college in 2009, the prevailing advice from experts was that young people needed to develop "hard skills," like computer science and engineering, that would give them a leg up in the job market. We were told that the STEM subjects—science, technology, engineering, and math—were the future, and that studying philosophy or art history or some other soon-to-be-obsolete subject would basically guarantee you a life of poverty and irrelevance.

This sneering attitude toward the humanities was reinforced by political and business leaders, who fretted that America wasn't equipping enough graduates with the right skills for a twenty-first-century economy. Marc Andreessen, the venture capitalist and Netscape cofounder, said at a 2012 tech conference that most English majors would "end up working in a shoe store." Vinod Khosla, the venture capitalist and Sun Microsystems cofounder, declared in a 2016 blog post that "little of the material taught in Liberal Arts programs today is relevant to the future." Even President Obama argued that the humanities were fading into irrelevance, saying in a 2014 speech, "I promise you, folks can make a lot more potentially with skilled manufacturing or the trades than they might with an art history degree."

Around the same time that STEM supremacists were preaching the value of hard skills, the concept of "lifehacking" came into vogue. The assumption behind this trend, which was especially popular among engineers in Silicon Valley, was that our bodies and minds could be optimized and improved, in the same way you'd speed up a sluggish computer. Life coaches and social media gurus gave sermons on personal productivity, and advised people to eliminate all waste and inefficiency from their daily lives. Sites like Lifehacker and Medium overflowed with tips for productive living—everything from dot-journaling to drinking Soylent—and we obsessed over the latest methods for getting more done. The unspoken message behind all of this advice was the same: *your humanity is a bug, not a feature.*

For many years, that message was essentially accurate, at least in a rational, economic sense. The industrial economies of the nineteenth and twentieth centuries required workers who could perform repetitive tasks at a high, consistent level, and individuality, in a factory setting, could be a liability. (Henry Ford's famous supposed lament about his workers— "Why is it every time I ask for a pair of hands, they come with a brain attached?"—echoed the feelings of many old-economy barons.) White-collar knowledge workers performed cognitive labor instead of manual labor, but they, too, often benefited from suppressing their humanity in the quest for peak performance.

But as I started reporting more on AI and automation, the message I heard from experts about the modern economy was, essentially, the exact opposite.

These experts said that, in a highly automated economy, the most valuable skills and abilities were the ones that could *distinguish* workers from machines. Rather than treating ourselves as pieces of biological hardware to be debugged and optimized, they said, we needed to develop the kinds of unique, human skills that machines couldn't replicate.

This conclusion made a certain kind of sense. And it tracked with other research I'd done, which found that the people who succeeded during periods of rapid technological change throughout history weren't always high-tech engineers and programmers. Often, they were the people who did the kind of low-tech, high-touch jobs that machines couldn't replicate.

During the Industrial Revolution of the eighteenth and nineteenth centuries, for example, there was a huge boom in factory work, but there was also a spike in demand for teachers, ministers, civil engineers, and other professionals who could serve the new, denser urban populations. During the manufacturing automation boom of the mid-twentieth century, as producing physical goods became cheaper and more efficient, a greater share of economic activity moved to fields like education and healthcare, where there weren't as many robots or fancy machines to do the work. And over the past few decades, as tech companies have taken over the economy, some of the fastest-growing occupations in America—massage

therapists, speech-language pathologists, animal caretakers—
have been decidedly analog.

As I explored these trends, it became clear that in order to
figure out a survival strategy for the future, I needed to start by
understanding where today's machines are weak, relative to
humans. So, I started asking experts one question:

*What can humans do much, much better than even our most
advanced AI?*

Surprising

The first thing I learned is that, in general, AI is better than
humans at operating in stable environments, with static, well-
defined rules and consistent inputs. On the other hand, humans are much better than AI at handling surprises, filling in
gaps, or operating in environments with poorly defined rules or
incomplete information.

This is why, for example, a computer can beat a human
grandmaster at chess, but would make for an extraordinarily
bad kindergarten teacher. It's why virtual assistants like Siri
and Alexa respond well to simple, structured questions that
draw from concrete data sets ("What's the weather in New
York next Tuesday?") but freeze up when confronted with
questions that require handling uncertainty or drawing infer-
ences from incomplete data ("What's the restaurant near
Gramercy Park with the really good burger?").

Even small surprises can trip up AI badly. In a 2018 experiment, a group of AI researchers took a deep neural network—
the kind of system that helps apps like Google Photos recognize
objects and faces inside your pictures—and trained it to recognize objects inside a photo of a living room. After being shown

millions of examples, the AI correctly identified the objects in the room: chair, person, books. Then the researchers introduced a surprise into the living room—a tiny image of an elephant—and ran the model again. This time, the results were a disaster. The AI labeled a chair a couch, mistook the elephant for a chair, and mislabeled other objects it had previously identified correctly. The presence of one anomaly hadn't just made the AI freeze up—it seemingly gave the AI a nervous breakdown, in which it appeared to forget everything it had ever learned.

This kind of thing doesn't happen to humans. When we see something unexpected, we do a double-take—we back up and reprocess the visual information, making different assumptions about what it might represent. But today's AIs can't do that. Since they have no holistic model of the world and how humans interact with it (what we might call "common sense"), most AIs depend on having lots of high-quality examples at their disposal.

There are types of AI that don't require a whole bunch of labeled data, such as unsupervised learning, a technique in which an algorithm is told to go out and hunt for patterns in a big, messy data set. And some types of AI are getting better at handling new situations. But they're still pretty far away from being able to navigate them with ease. Which means that humans who are good at handling the unexpected—who are cool in a crisis, who like dealing with messy problems and novel scenarios, and who can move forward even in the absence of a concrete plan—still have an advantage.

This bodes well for people whose jobs involve constant change. Occupational therapists, police detectives, emergency room nurses—these are the kinds of jobs that rarely look exactly the same from day to day, and in which truly repetitive tasks are somewhat rare.

It bodes poorly, however, for jobs that are very structured and highly repetitive, such as data entry clerks, loan underwriters, or tax auditors. One AI expert put it to me this way: if you could write a user's manual for your job, give it to someone else, and that person could learn to do your job as well as you in a month or less, you're probably going to be replaced by a machine.

Social

The second thing I learned is that while AI is really good at meeting many of our material needs, humans are much better at meeting our social needs.

There are some areas of life in which only outcomes matter. We don't really care if our subway car is driven by a person or a computer, as long as it's safe and efficient and gets us to our destination. Few people would object to a robot handling their packages in a warehouse, as long as the packages arrived on time and intact.

But many things in life are not bloodless exchanges of currency for goods and services.

Humans are social beings. We like feeling connected to one another and having meaningful interactions with the people around us. We care deeply about our social status, and what other people think of us. And many of the choices we make every day—even the seemingly mundane ones, like the food we eat or the clothes we wear—are actually deeply related to our identities, our values, and our need for human connection.

What that means, practically speaking, is that jobs that tap into our social desires—bartenders, hair stylists, flight attendants, mental health workers—will be hard to automate. And people who are skilled at creating social and emotional experiences will be better positioned for the future than people whose primary skill is making or doing things efficiently.

The value of emotional intelligence is already obvious in jobs like nursing, ministry, and teaching. But as AI and automation enter more fields, making people feel connected and socially fulfilled will become a high-value skill in those fields, too. Good lawyers will become more like legal therapists—creating trust with clients and helping solve their problems, rather than simply writing briefs and doing research. Doctors will be sought-after based on how they interact with patients, rather than how well they know the latest treatment protocols. Successful programmers won't just be isolated geniuses pecking out lines of code; they'll be people who can lead teams, think strategically, and explain complicated technical concepts to non-programmers.

It's not that technical skills, or basic competence, will cease to matter in the age of AI and automation. It's just that when machines can do many of the basic, repetitive functions of our jobs as well or better than we can, what's left for us will be the social and emotional parts.

You can already see this shift happening in many industries. The travel agents who have survived the rise of Kayak, Expedia, and Orbitz have largely done it by focusing on creating unique experiences for travelers—wilderness adventures, cooking classes, authentic homestays—rather than just finding them good deals on hotel rooms. In advertising, where most of the day-to-day work of media buying can now be done by programmatic algorithms, many of the remaining jobs are in areas like creative, client services, and influencer marketing, where understanding human desires and being able to work closely with other people is a key element of what makes you successful.

A good general rule is that jobs that make people *feel* things are much safer than jobs that simply involve *making* or *doing* things.

Scarce

The third thing I learned is that AI is much better than humans at big work—work that involves large data sets, huge numbers of users, or global-scale systems. If you need to produce a million of something or spot the patterns in a hundred thousand data points, you're probably looking at a job that is already done by a machine, or soon will be.

On the other hand, humans are much better than AI at work that involves unusual combinations of skills, high-stakes situations, or extraordinary talent.

I call this type of work "scarce," but not because there are only a few of these jobs to go around. Instead, this is work that will be either impractical or socially unacceptable to automate because it isn't needed on a constant, predictable basis.

Most AI is built to solve a single problem, and fails when you ask it to do something else. An AI that learns to make video recommendations at a world-class level generally can't be repurposed to audit financial statements or filter email spam. And so far, AI has fared poorly at what is called "transfer learning"—using information gained while solving one problem to do something else. (The exceptions to this rule are deep learning algorithms like AlphaZero, the AI built by Google's DeepMind, which recently taught itself to play chess and Go at a world-class level in a matter of hours by playing against itself millions of times. But even AlphaZero is limited to the world of games—it couldn't, for example, unclog a sink.)

Humans, by contrast, are great connectors. We spot a problem in one area of our life and use information we learned doing something completely different to fix it. We take a piece of advice a teacher gave us in middle school and apply it to life situations that arise decades later. We remix ideas, blend genres, and hold vast amounts of random, disparate information in our heads, ready to be mashed together at a moment's notice.

Maria Popova, the creator of the Brain Pickings blog, calls this trait "combinatorial creativity." She writes that many of history's great breakthroughs have been generated not by hyper-specialization, but by combining insights from two or more different fields. She cites Albert Einstein, who said that playing the violin helped connect different parts of his brain while working on physics problems, and Vladimir Nabokov, the Russian-born novelist, who credited his hobby as a butterfly collector with making his writing more detailed and precise.

For now, combinatorial creativity is a uniquely human skill. Which means that people with unusual combinations of skills—like a zoologist with a math degree, or a graphic designer who knows everything there is to know about folk music—will have an upper hand against AI.

Another type of scarce work that will be hard to automate is work that involves rare or high-stakes situations with low fault tolerance.

Most AI learns in an iterative way—that is, it repeats a task over and over again, getting it a little more right each time. But in the real world, we don't always have time to run a thousand tests, and we know, intuitively, that there are things that are too important to entrust to machines. When we call 911, we want a human, not an automated answering service, to take our call. When an engaged couple wants to make sure every detail of their wedding goes off without a hitch, they hire a wedding planner, not an automated logistics firm. When babies are born, we want a human doctor to be present in case something goes wrong, even if a virtual obstetrician could perform that work 99 percent of the time.

Scarce jobs also include jobs that require human accountability or emotional catharsis. When our health insurance company wrongfully denies a covered claim, or an Airbnb guest trashes our house, we don't want to fill out a form on a web portal—we want to complain to a human, and get our issues resolved.

The final kind of scarce work that is almost certainly safe from automation is work that requires extraordinary talent. World-class athletes, prize-winning chefs, and people with standout acting or singing abilities all fit into this category. Basically, if someone would pay to watch you do your job, it's probably safe.

The reason these jobs are unlikely to be automated is less about the limitations of technology, and more about our own intrinsic needs. No matter how good AI gets, humans still want role models, and we want to be inspired by human greatness. This is why we cheer for Olympic swimmers, even though speedboats go faster. It's why we'll pay to see our favorite band perform live, even when we could stream their music for free at home. We like bearing witness to human greatness, and we don't yet accept machine substitutes.

As I thought about these conclusions, I realized that since my first automation scare, I'd unwittingly made my own job more surprising, social, and scarce.

I'd stopped doing formulaic corporate earnings stories and started writing stories that required more creativity and revealed more of my personality—stories that made people feel things, rather than simply conveying information. I got off the Wall Street beat and started writing about technology, and spent months digging into fringe internet communities, which gave me a relatively scarce body of knowledge I could use to advance my career. I also started broadening my journalistic tool kit—producing and co-hosting a TV show, making a podcast—which gave me more possible combinations of skills I could bring to bear on a project.

As I made these changes, I started noticing examples of other people who had succeeded by making themselves more surprising, social, and scarce everywhere I looked.

Take my accountant, for example. His name is Rus Garofalo, and he does my taxes every April. Rus is not a typical tax preparer. He's a former stand-up comedian, and he brings his comedic sensibility to his work. (The name of his company is Brass Taxes—get it?)

Rus knew that in the age of TurboTax, the only way to survive as a human accountant is to bring something to the table besides tax expertise. So, he hired a bunch of other funny, personable accountants. He offered to pay for improv comedy classes for all of them. And he started looking for creative clients, like actors and artists, who tended to have more complex tax returns, and who would appreciate having a real person guide them through the process.

Technically, I should be worried about Rus, because tax preparation is a very automation-prone occupation. (In fact, according to a recent Oxford University study, it has a 99 percent chance of being automated.) But I'm not worried about Rus, because he's figured out a way to turn his services from a routine transaction into a surprising, social, and scarce experience that people, including me, are happy to pay for. I asked Rus if he'd done this on purpose, knowing that robots were coming for his job. And he said, in so many words, that he had.

"A lot of tax preparers want you to drop off your papers, go away, and send them a check for $400—that's ideal market efficiency, and that's why TurboTax killed them," he told me. "Our defining value we add is the conversation we have with you."

I also found examples of businesses that hadn't explicitly been threatened by automation, but whose surprising, social, and scarce approach to their work had helped them weather other types of storms.

Like Marcus Books. Marcus Books is an independent, Black-owned bookstore in my hometown of Oakland. It's the oldest Black-owned bookstore in America, and an amazing place with a sixty-year legacy of introducing Oaklanders to the works of incredible Black authors like Toni Morrison and Maya Angelou.

But maybe the most amazing thing about Marcus Books is that it's still alive. There are very, very few independent bookstores left in the Bay Area, and almost no Black-owned bookstores have been able to survive the onslaught of Amazon and the internet.

So how did they do it? It's not because they have the lowest prices, or the slickest e-commerce system. It's because Marcus Books isn't just a bookstore. It's a community hub, a store filled with kind employees who have actually read the books they recommend, and a safe place where Black customers know they won't get followed around or patted down by a security guard. Above all, it's a place with what its co-owner, Blanche Richardson, described to me as "good vibes."

In early 2020, when the Covid-19 pandemic hit the Bay Area, Marcus Books was forced to temporarily close its doors. And like a lot of businesses, it faced an uncertain future. But its community rallied behind it, starting a GoFundMe page and raising money to keep the store in business.

Then, in May, George Floyd, an unarmed Black man, was killed by police in Minneapolis. Protests filled the streets of America's cities, and orders started pouring into Marcus Books from all over the country from people who wanted to support its mission. It's selling five times as many books as it was before the pandemic, and its GoFundMe donations ballooned to $260,000, more than enough to keep the store afloat.

Marcus Books is not a high-tech operation. (In fact, until recently, you couldn't even order books on its website.) And if all you were looking for was a large selection of cheap books, you'd probably order them from Amazon. But the store has something that turned out to be much more valuable than a website—an authentic connection to a community that had their back when trouble hit.

What has helped Marcus Books survive for sixty years, even as the world has changed around it, is that it is still selling books in a surprising, social, and scarce way. It has kept its humanity front and center, and as a result, it has made itself irreplaceable.

William Lovett also survived by becoming more surprising, social, and scarce.

After arriving in London in 1821, Lovett talked his way into a job at a carpentry shop, whose owner offered to teach him the trade of cabinetmaking—a task that, unlike rope making, didn't lend itself to mass production.

With a stable job in hand, Lovett began pursuing more intellectual interests. He joined a men's group that met regularly in an old meat market, where they spent hours discussing politics, theology, and classic literature.

"My mind seemed to be awakened to a new mental existence," he wrote in his autobiography. "New feelings, hopes, and aspirations sprang up inside me, and every spare moment was devoted to the acquisition of some kind of useful knowledge."

Soon, Lovett became active in labor organizing—a deeply interpersonal endeavor, and one that made him a sought-after ally as British workers fought for expanded rights and protections. He helped lead the Chartists, a working-class reform group, and became involved in education, advocating for a model of schooling that emphasized teaching human qualities like tolerance, love, and compassion rather than rote skills and tradecraft.

Education, Lovett wrote, "must comprise the judicious development and training of ALL the human faculties, and not, as is generally supposed, the mere teaching of 'reading, writing, and arithmetic,' or even the superior attainments of our colleges, Greek, Latin, and polite literature."

Lovett never got rich or famous. You won't find libraries or university buildings festooned with his name. But he did something that, in its own way, was remarkable. In a time of incredible technological change, he figured out how to stay ahead of the curve by centering his humanity in his work. He realized that his intellect, his relationships, and his moral courage made him much more valuable than a machine and acted accordingly.

As a result, he was able to build a life with meaning and purpose. He became futureproof. And as far as we know, he never made rope again.

Rule 2

Resist Machine Drift

The main business of humanity is to do a good job of being human beings, not to serve as appendages to machines, institutions, and systems.

—KURT VONNEGUT

Before we go any further, do you mind if I ask you a few personal questions?

Lately, have certain parts of your life felt a little . . . predictable?

Do you and your friends mostly watch the same TV shows, read the same books, and listen to the same podcasts?

Could a stranger predict your exact tastes in clothing, food, and politics knowing only your age, gender, ethnicity, and zip code?

Have you caught yourself coasting on mental autopilot—saying the obvious things, repeating the same activities, going through the motions without any variety or serendipity—for weeks or months at a time?

Do you ever look back at pictures or videos from years ago and feel like you're seeing not just a skinnier or fresher-faced version of you, but a more *surprising* version of you, one who thought more independently, engaged more deeply with ideas, and ventured further out of the mainstream?

I'll go first. I feel all of these things, all of the time. And I don't just think it's the haze of nostalgia. I think it's about the machines.

So far in this book, we've talked mainly about external forms of automation—industrial robots, machine learning algorithms, back-office AI software. But there is a kind of *internalized* automation taking place inside many of us that, in some ways, is much more dangerous. This kind of automation burrows into our brains and affects our inner lives— changing how we think, what we desire, whom we trust. And when it goes haywire, it can cost us much more than a job.

I've seen lots of examples of this kind of automation in the last few years, as I've covered social media for the *Times*. I've interviewed followers of online extremist movements like QAnon, and I've seen how the algorithms and incentives of social media can turn normal, well-adjusted people into unhinged conspiracy theorists. In an audio series I helped report, *Rabbit Hole,* I examined the ways platforms like YouTube and Facebook have been engineered to use AI to lure users into personalized niches filled with exactly the content that is most likely to keep their attention—and how, often, that means showing them a version of reality that is more extreme, more divisive, and less fact-based than the world outside their screens.

We don't often talk about AI in the same breath as things like social media misinformation and online radicalization. But they are closely related. AI is what makes these platforms so addictive. And their ability to pinpoint exactly what will keep us clicking, watching, and scrolling is, ultimately, what creates the potential for manipulation.

I'll admit that I've let machines run my life to an embarrassing extent over the years. I've used AI assistants to manage my calendar, bought robot vacuums and Wi-Fi-connected thermostats to keep my house clean and temperature-controlled, and subscribed to wardrobe-in-a-box services that use fancy algorithms to figure out which clothes will look best on my body type. At work, I've used canned email snippets and relied on those auto-generated Gmail replies to save time. ("Yep!" "Sure, that works!" "No, I can't.") For years, I mostly went with the algorithmic flow—ordering the stuff Amazon suggested for me, playing auto-generated Spotify playlists, and watching the shows Netflix recommended.

For a long time, all of this lifestyle automation seemed harmless. But eventually, I began feeling that surrendering my daily decisions to machines wasn't making me happier or more productive. Instead, it was turning me into a different person—a shallower one, with more fixed routines and patterns of thought, and an almost robotic predictability in my daily life.

I started calling this feeling "machine drift," and I first noticed it happening to me a few years ago.

At the time, I was an editor at a digital news site, and my job included making sure my section hit its monthly traffic targets. Toward the end of every month, if we were lagging behind our goals, I'd scramble to put up viral stories that might reel in big traffic from Facebook or Google. I got pretty good at it. One of my posts, a warmed-over summary of a Reddit thread, hit the Facebook click jackpot and got several million views. Another story, which brought in several more million views, was a four-sentence-long post titled "Ann Coulter did a bad tweet."

These end-of-the-month posts did their job, but whenever I wrote them, I started to feel less like a journalist and more like a factory worker shoveling coal into a furnace. I wasn't doing anything novel or creative. I was just feeding a couple of algorithms what they wanted—and in the process, I was becoming a kind of algorithm myself.

I experienced machine drift outside of work, too. I felt myself becoming sharper and more politically polarized, and many of my weaker preferences calcified into strong, bedrock beliefs. More of my thoughts arrived as Twitter-sized witticisms, and I had a harder time listening to opposing points of view with an open mind.

Once I connected these feelings to my use of technology, I started second-guessing myself. Did I actually like the leather sneakers I bought on Amazon, or did I just trust the algorithm more than my own fashion sense? Was I actually mad at the venture capitalist whose dumb tweet crossed my Twitter timeline, or was I just joining a dogpile because I knew Twitter's algorithm would reward my snarky joke with likes and retweets? Did I really like cooking, or did I just like the way my Instagram photos of home-cooked meals made me look like a balanced, well-adjusted adult?

How many of my beliefs and preferences were actually mine, I wondered, and how many had been put there by machines?

• • •

In 1990, two scientists at Xerox PARC, the Palo Alto–based research lab, came up with a way to solve the annoying problem of email overload. Email was still a new technology, and inboxes at Xerox PARC were overflowing with off-topic and unnecessary messages. Researchers were spending hours a day reading and deleting emails from various news groups they subscribed to, and it was interfering with their work.

One day, Doug Terry, a junior researcher at the lab, had an idea. What if, instead of presenting your emails in chronological order, an email program could rank them in order of importance? And what if the news you saw was determined, in part, by what other people had already read and liked? He recruited another engineer, David Nichols, and the two of them began working on a program called "The Information Tapestry"—or "Tapestry," for short—that could help them bring order to their inboxes.

Their first task was to create an automatic ranking system for regular, person-to-person emails. They programmed a series of "appraisers"—algorithms that would scan incoming emails and assign each message a priority score based on factors like the sender's name, the subject line, and the number of other recipients. Emails from Terry's boss that were sent only to Terry would receive a priority score of 99—the highest score possible—and would always appear at the top of his inbox. Below those might be emails containing medium-important keywords like "Apple" (one of Xerox PARC's biggest competitors) or "baseball" (Terry's favorite sport). Emails from unknown senders with no relevant keywords would receive low scores and would appear near the bottom of the inbox.

The second step was to figure out a way to sort the hundreds of impersonal emails that flooded into their inboxes every day from news groups and clipping services. Terry and Nichols came up with a system they called "collaborative filtering," which would allow users to prioritize messages based on other users' recommendations—in effect, turning their colleagues into a filtering algorithm.

Collaborative filtering worked by adding two buttons to the end of each news group message. One said "Like it!" and the other said "Hate it!" Depending on which button a user clicked, that story would appear higher or lower in the inboxes of other users. Users could personalize their filters by subscribing to recommendations from specific people, on specific topics, or in specific groups, and they could chain together multiple filters to create a customized recommender. ("Show me articles recommended by John Smith on the topic of Yankees in the comp.misc.baseball newsgroup.")

Terry and Nichols, along with two other researchers they recruited, spent the next six or so months building Tapestry. Then, they released it to their colleagues. Several dozen people signed up. And the recommendation engine was born.

Today, the world runs on recommendation engines. Right now, as you read this, billions of people all over the world are using algorithmically generated recommendations to help them decide which clothes to wear, which trips to take, which jobs to apply for, which groceries to buy, which plumbers to hire, which stocks to invest in, which TV shows to watch, which restaurants to patronize, which music to listen to, and which potential romantic partners to date. Our entire information ecosystem is wrapped around the recommendation engines that power social media platforms like Facebook, Twitter, and YouTube—all of which rely on algorithms to tell us which voices matter, which stories are important, and what deserves our attention. Our politics, our culture, even our interpersonal relationships are bound up in the recommendations these systems make, and the gymnastic ways people try to reverse-engineer and game them.

The injection of algorithmic recommendations into every facet of modern life has gone mostly unnoticed, and yet, if we consider how many of our daily decisions we outsource to machines, it's hard not to think that a historic, species-level transformation is taking place.

"Recommendation engines increasingly shape who people are, what they desire, and who they want to become," writes Michael Schrage, an MIT research fellow and author of a book about recommendation engines.

"The future of the self," he adds, "is the future of recommendation."

Modern recommendation systems are orders of magnitude more powerful than the one Doug Terry and Dave Nichols developed to sift through their email inboxes. Today's tech companies have access to huge amounts of computing power that allows them to generate detailed models of user behavior, and machine learning techniques that let them discover patterns in enormous data sets—studying the online shopping behavior of a hundred million people to find out, for example, that people who buy a certain brand of dog food are statistically more likely to vote Republican.

The other big difference is that while recommender systems of the past were designed to save us time, many of today's recommenders are designed to take time from us. Facebook, Instagram, YouTube, Netflix, Spotify, and even *The New York Times* use recommenders to personalize users' feeds, showing them what the machines believe will keep them engaged for as long as possible.

These algorithms can be shockingly effective. YouTube has said that recommendations are responsible for more than 70 percent of all time spent on the site. It has been estimated that 30 percent of Amazon page views come from recommendations—a figure that could translate to tens of billions of dollars in annual sales. Spotify's algorithmically generated Discover Weekly playlists have become music industry hitmakers in their own right and reportedly account for more than half the monthly streams of more than eight thousand artists. Netflix has said that 80 percent of the movies viewed on its service come from recommendations and has estimated that its recommendations save the company $1 billion a year.

The psychological power of recommendations was made clear in a 2018 study led by Gediminas Adomavicius, a professor at the University of Minnesota.

The study consisted of three experiments. In the first, participants were given a list of songs, with one-to-five-star ratings attached to each song. (These ratings were randomly assigned, but participants were told that they were based on their music preferences.) Participants were given the chance to listen to a short sample of each song, if they wanted; then they were asked how much they would be willing to pay for each song.

In the second experiment, participants were given real song recommendations generated by an algorithm like the ones used by streaming music services like Pandora and Spotify. But researchers manipulated the ratings beforehand, intentionally adding stars to some songs and removing stars from others. Similar to the first experiment, participants had the option of listening to short clips before placing a value on each song.

In the third experiment, researchers again gave each song a random rating, but this time, participants were required to listen to the songs in their entirety before valuing them.

The results of the first two experiments weren't surprising. Participants trusted the star ratings, even when they weren't truly personalized, and placed a higher value on songs that had been assigned a higher rating.

The results of the third experiment, however, took the researchers aback. They had expected that forcing participants to listen to each song before valuing it would neutralize the effects of the star ratings. (Actually having listened to a song, they thought, was a much better indicator of whether or not you liked it than knowing what an algorithm had said about it.) But participants still offered significantly more money for higher-rated songs. In other words, the machines' randomized preferences overrode their own experiences.

"Consumers don't just prefer what they have experienced and know they enjoy," the researchers wrote. "They prefer what the system said they would like."

At their best, recommenders are a beautiful and empowering form of consumer leverage—a way to put powerful machines to work as our personal concierges, sorting through the vast expanse of the internet to create an experience tailored to our preferences.

At their worst, they're more like pushy salespeople—

shoving options we don't want in front of us, playing psychological games, hoping we'll relent. With recommenders, we're still technically in control. (We are, after all, humans with agency and free will.) But the force these systems exert on us is not always the nudge of a friendly suggestion. Often, they coerce us in their preferred direction by arranging our choices for us—making desirable options easier or more prominent, while burying undesirable options several clicks deep in a menu. Many recommenders come attached to friction-reducing features like autoplay and one-tap checkout, all of which are designed to speed us to a decision before we can stop and consider whether the machine's preferences actually match our own.

The fact that machines can shape our preferences is not a secret in Silicon Valley. In fact, there is an entire subdiscipline of product design, called "choice architecture," that uses subtle design elements to change the things users click, buy, and pay attention to.

Some choice architecture can be helpful—it's good, for example, that Yelp shows you nearby and well-reviewed restaurants by default and doesn't make you sort through an alphabetical list of all the restaurants in your city. But choice architecture can also steer our attention to things we don't want, that aren't beneficial for us, and that we wouldn't have sought out on our own.

The technology scholar Christian Sandvig calls this "corrupt personalization," and it's never more apparent than when companies put their own thumbs on the scale. Just by tweaking its algorithms, Netflix can steer users to its original shows,

Amazon can steer users to its house brands, and Apple can recommend its own apps in the App Store, even when other apps might be preferable.

The power to change users' preferences at scale has made some technologists uncomfortable. Rachel Schutt, a data scientist, said as much in a 2012 interview with the *Times*: "Models do not just predict," she said, "but they can make things happen." A former product manager at Facebook went even further, telling BuzzFeed News that Facebook's recommendation algorithms amounted to an attempt to "reprogram humans."

"It's hard to believe that you could get humans to override all of their values that they came in with," the former Facebook employee said. "But with a system like this, you can. I found that a bit terrifying."

The French researcher Camille Roth divides digital recommendations systems into two camps: "read our minds" algorithms, which aim to apply our existing preferences to new information, and "change our minds" algorithms, which attempt to transform our preferences into different ones, or create preferences where we previously had none.

For years, most recommenders were mind-reading algorithms—they tried to predict what you wanted to see and show it to you. But in recent years, tech companies have figured out that mind-changing algorithms are a lucrative prospect. Targeted advertising—the business that has made Google and Facebook two of the most valuable companies in the world—combines both "read our minds" and "change our minds" technology by analyzing data to guess users' preferences (the "targeted" part), then allowing advertisers to pay to attempt to change their minds (the "advertising" part).

Today's recommender algorithms are so powerful, and so deeply embedded in our systems, that they often function more like *decider* algorithms. By ranking certain information more highly, or prioritizing choices in a certain way, they can create the illusion of free will, while actually steering users down a path toward their preferred outcome.

What makes machine drift so dangerous is that, at a time when we most need our human faculties to help guide us, these algorithms are actively eroding the parts of ourselves that make us most human: our ability to change course, to pursue difficult goals, to make unpopular choices that cut against the grain. They're discouraging us from building the kind of personal autonomy that will protect us in the age of AI and automation, by allowing us to think and act for ourselves. And they're doing it under the guise of helping us.

In a 2017 paper about the history of Amazon's recommendation algorithms, Amazon engineer Brent Smith and Microsoft data scientist Greg Linden sketched out a vision of the AI-driven future that feels, to me, both deeply dystopian and very, very plausible.

"Every interaction should reflect who you are and what you like, and help you find what other people like you have already discovered," they wrote. "It should feel hollow and pathetic when you see something that's obviously not you; do you not know me by now?"

"Getting to this point," they continued, "requires a new way of thinking about recommendations. There shouldn't be recommendation features and recommendation engines. Instead, understanding you, others, and what's available should be part of every interaction."

Every interaction. It is not enough to accompany us to the store, whispering into our ears about which brand of toothpaste or toilet paper we should buy. In the eyes of engineers and executives who use recommendation algorithms to steer our choices, all of our actions must be part of the machine's model. There is no space, in this vision of the automated future, for developing new tastes, or starting over with a clean slate. Who you are is who the machines think you are, which is also who they want you to be.

If recommendation algorithms are one ingredient in machine drift, the other ingredient is what is known, in Silicon Valley–speak, as "frictionless design."

For a modern technologist, there is no greater enemy than friction—not the literal, physics-class kind, but the metaphorical kind that occurs whenever a user encounters an unnecessary delay or inefficiency on their way to accomplishing a given task. Tech companies have spent decades profitably stripping friction out of our lives, making it easier to hail taxis, order household goods, or pay for things at a store, and frictionless design has become a kind of religious tenet for aspiring tech moguls. Brenden Mulligan, a tech entrepreneur, summarized Silicon Valley's anti-friction ethos in a TechCrunch essay.

"If your users feel friction using or signing up for your service, you have a problem," Mulligan wrote. "Sometimes it's unavoidable, but you should do everything in your power to remove as much friction as possible."

I first encountered the idea of frictionless design in 2011, when Facebook CEO Mark Zuckerberg announced that Face-

book was introducing a new feature called "frictionless sharing." The feature—which allowed certain apps, like Netflix
and Spotify, to post directly to users' feeds, rather than having
to ask for permission each time—was a failure, and Facebook
killed it fairly quickly. But the idea of a "frictionless" product
captured Silicon Valley's imagination. Uber, Square, and other
tech companies committed themselves to frictionless design.
Jeff Bezos, Amazon's founder, articulated the strategic benefits
of reducing friction in a 2011 letter to investors.

"When you reduce friction, make something easy, people
do more of it," Bezos wrote.

Many types of friction-reducing design are unequivocally
good. We don't want friction at the doctor's office or the DMV.
There is nothing virtuous or romantic about expending unnecessary effort booking a flight, filing an insurance claim, or applying for unemployment benefits. There are also still many,
many Americans with far too much friction in their daily lives,
and the minor inconveniences privileged white men like me
sometimes refer to as "friction"—having to fax a form to a government office, for example—are often overstated.

But Silicon Valley's friction fighting has come at a price.
Some of it is related to where friction goes, once it has disappeared from a consumer's phone or computer screen. Often
"eliminating friction" in a tech product simply means trans-
ferring the burden to a low-paid worker. Amazon's all-out
attempts to reduce customer friction have put additional pressures on its warehouse workers. Uber drivers lost out on millions of dollars in customer tips because former Uber CEO
Travis Kalanick thought that giving riders a tipping option
inside the app would create unnecessary friction. (After Ka-

lanick's ouster, the company came to its senses and added the feature.)

The biggest problem with frictionless systems, though, is what they do to our autonomy. Just like recommendation algorithms, they pull us into the middle of the bell curve—training us to pick the most popular option, the most probable outcome, the path of least resistance. They rarely prod us to do the hard, counterintuitive thing, or pause to scrutinize our own impulses. And by reinforcing what the technology critic Tim Wu calls "the tyranny of convenience"—the idea that the best solution is always the easiest one—they can cause us to overlook things we might value more in the long term, like trying new experiences, or overcoming tough obstacles.

It's not surprising that recommendation systems and frictionless design have caught on, given the way they can strip complexity out of our chaotic, fast-moving lives. And, to reiterate, not all personalized recommendations or frictionless apps are bad.

But we have to be careful about giving too much of ourselves to our tools. Because the philosophy that gives rise to machine drift is, fundamentally, nihilism. It's an attempt to persuade us that there is nothing important about us that cannot be quantified or reduced to a series of data points, or any inner life worth protecting from machine influence. Recommendation engines and frictionless products offer us their help, but their ultimate goal is surrender—a swimmer caught in a riptide, who gets tired of fighting the current and simply decides to float.

• • •

As a first step to resisting machine drift, I recommend taking an inventory of your own preferences. Keep track of all of the choices you make in a day and try to determine which of those choices are truly yours, and which are fundamentally shaped by a machine's instructions or suggestions. Do you buy the same brand of dog food every month because it's what Amazon recommended to you, or because your dog actually likes it? Does the route you take to work reflect your idea of a good commute, or Google Maps' idea of the optimally efficient journey? Would you take that hike, wear that jacket, or loudly state that political position if there were no likes, views, or retweets hanging in the balance—if it was only about you, who you are, and what would bring you the most pleasure and fulfillment?

Once you've teased apart your own preferences, values, and priorities, write them down. What hobbies and activities do you actually like doing? What political and spiritual beliefs would you still have in a vacuum, even if nobody would ever know you believed them? Which relationships actually enrich your life? Keep that list handy. Put it up on the wall, if you want. This is, to a first approximation, a blueprint of your core self—and it will remain a useful reference point.

Another way to resist machine drift is to implement what I call "human hour." Every weekday, around the same time—for me, it's usually five to six P.M.—I try to spend at least an hour away from screens, doing something I genuinely enjoy doing, like playing tennis, cooking, or taking my dog for a run. Choosing purely optional activities is key—I don't cross items off my to-do list during human hour or take care of household chores. The point is to spend an hour a day getting back in touch with my own needs and priorities by doing things that make me feel more human, while escaping the web of incentives and invisible forces that tug on me throughout the rest of the day.

In the interest of resisting machine drift, I've also started adding a little more friction to my daily routine. Instead of ordering a power drill on Amazon, I drive to the local hardware store. I take the extra two minutes to warm up milk for my morning coffee, rather than pouring it in cold. I read the print newspaper on the weekends instead of scrolling through Twitter headlines. When I commute to my office, I take a longer, more scenic route that adds about fifteen minutes to the trip but makes it much more pleasant.

To be clear, these are *extremely* minor inconveniences, and I'm beyond fortunate to have the time and flexibility to take them on willingly. Many people work much harder than I do, in conditions far less hospitable than mine, and they need every bit of convenience they can get. I hope tech engineers and product designers will find ways to reduce friction for vulnerable and marginalized people, rather than shaving slivers of discomfort off the lives of the already-convenienced.

But for those of us privileged enough to choose our own tempos, a lifestyle with slightly more friction and autonomy can be gratifying. After all, few of our happiest moments or proudest achievements are the result of letting algorithms make our decisions for us. The mountains climbed and marathons finished and children successfully parented—they all result from intentional choices to do *more* than is strictly necessary. What is rewarding is often hard, and hard is the enemy of the machine.

• • •

Recently, I called Doug Terry, the Xerox PARC engineer who came up with Tapestry, the first algorithmic recommender system, nearly three decades ago. Terry, who is sixty-two, works at Amazon now, and after reminiscing about the early days of Tapestry, I asked what he thought of the recommendation engines that power services like Facebook, YouTube, and Netflix.

"I don't think there's any comparison," he said. "We just had a little simple system, and nowadays there's trillions and trillions of feeds for billions of people—just the scale and complexity and everything is different."

Terry couldn't have known, back when he came up with the concept of filtering his news diet based on his colleagues' recommendations, that the same technology would fuel the rise of multibillion-dollar tech giants and fundamentally change our global information ecosystem. And as I told him about all my concerns—my worries that algorithms designed to capture our preferences were actually distorting those preferences, my anxieties about machine drift, my sources who had been radicalized by their social media recommendations—he seemed troubled.

In the old days, he said, recommendations were simply that—recommendations. But now, they were playing a more determinative role. He described it as a "snowball effect," in which people are shown more and more things that align with the algorithm's determination of their interests, ultimately limiting their view of the world to the things they're already comfortable seeing.

"I think one of the challenges is getting people to get out of their comfort zone," he said. "Recommenders do the opposite—they narrow your horizon."

Rule 3

Demote Your Devices

Computers make excellent and efficient servants, but I have no wish to serve under them.

—MR. SPOCK, *Star Trek*

I remember the exact moment I realized that I hated my phone.

It was December 2018, a few days before Christmas. I was at a theater in Manhattan with my wife and a few friends, watching a performance by the world-class Alvin Ailey American Dance Theater. We'd been lucky to score pretty good seats, and I'd been looking forward to the show for weeks.

Midway through the first act, I felt a buzz in my pocket. I ignored it. A few minutes later, it buzzed again. *Hmm,* I thought. *Maybe the Instagram I posted earlier today is blowing up. Or what if I'm getting furious emails from my editor?* I tried to put it out of my mind and focus on the leaping dancers in front of me. But my imagination kept wandering. *What if my apartment is on fire? What if I accidentally posted something*

*controversial on Twitter, and President Trump is, at this very mo-
ment, calling me a "fake news moron from the New York Slimes?"*

I decided I couldn't wait to find out. I silently mouthed
"bathroom" to my wife, scooted past her and down the aisle,
rushed to the bathroom, entered a stall, and whipped out my
phone.

What I saw was, basically, nothing. A few unimportant
emails, a text from my pharmacy, an Instagram comment or
two. But the lack of urgent notifications didn't send me rush-
ing back to my seat. Instead, I stood in the stall, fully clothed,
for a good fifteen minutes while I checked Twitter and Facebook and caught up on everything I'd missed. When I emerged
from my daze and left the bathroom to return to my seat, I saw
a sea of other people flooding toward me. Intermission. I'd
missed the rest of the act.

A wave of shame washed over me as I realized that I hadn't
just deprived myself of seeing some spectacular dancing, but
that I'd done it for the dumbest possible reason. Presented with
the opportunity to experience something truly memorable, surrounded by people I loved, I'd instead chosen to lock myself in
a bathroom stall and flick at a screen, looking for a cheap dopamine hit. And I'd done it almost autonomically, as if some invisible force I was powerless to resist had been tugging at my brain.

When I found my wife, she asked where I'd been, and if
everything was okay.

"Something came up," I lied.

I got my first smartphone in 2006, during my freshman year of
college. It was a BlackBerry Pearl, a gray rectangular brick

with a white chiclet set in the middle of the keyboard. I was obsessed with it, and spent hours a day emailing, playing Brick Breaker, and coming up with clever BBMs—BlackBerry's proprietary text messages, which could only be sent to other BlackBerry owners and, therefore, became a status symbol among nerds on campus.

In a world of simple flip phones, having a BlackBerry was a superpower—an Alexandrian library in my pocket, ready at a moment's notice to look up any fact, settle any dispute, or communicate with anyone I'd ever met. It was less like getting a cool new gadget than gaining a kind of ambient hyperawareness, assembled via a constant trickle of real-time updates.

I assumed that eventually, the novelty would wear off. But it never did. Instead, I dug in deeper. When the iPhone was released in 2007, I lined up to get one. I started a Twitter account and set up an RSS reader. I formed group texts and got news alerts pushed to my home screen. My phone time kept creeping up—first to three or four hours a day, then to six or seven. Most nights, I slept with it inches from my head.

Until recently, my phone usage didn't feel particularly dysfunctional. But a year or two ago, I crossed the line into problem territory. Social media was increasingly making me irritated and angry. My twitchy attention span, ground down by years of push notifications and breaking news alerts, made it difficult for me to read books, watch full-length movies, or carry on extended conversations with my friends. I felt myself pulling back from the world, and my offline life began to seem grainy and sepia-toned, in comparison to the dynamic, high-resolution universe in my pocket.

For months, I tried to kick my phone habit by uninstalling Twitter and Facebook, turning my screen grayscale, or shuffling apps into hard-to-access folders. But nothing worked. My screen time kept rising, and my phone kept interfering with my life. One night, I got a notification telling me that my iPhone now had a screen-time tracking figure. Did I want to know my statistics? I did not, but it told me anyway. My daily average, it said, was nearly six hours. My highest one-day total: eight hours and twenty-eight minutes.

For years, I believed that my devices were expanding my awareness, enriching my social life, and extending my humanity in new directions. But I realized—first slowly, then, in the bathroom of the Alvin Ailey performance, all at once—that I was less a user of these devices than a servant to them. Every day, I paid attention to what my phone told me was important, let its beeps and buzzes determine my agenda, and absorbed its priorities as my own.

My phone had once been my trusted assistant. But at some point, it got a promotion and became my demanding, hard-driving nightmare of a boss.

You might not have a phone problem as severe as mine. But I'm willing to bet that, at some point in the past several years, you've found yourself checking your phone more often than you'd like or missed something important because you were too busy scrolling mindlessly through your Facebook or Twitter feed.

My goal isn't to make you feel guilty or scold you about phone addiction. I just want to encourage you to examine your relationship to your devices—which are, after all, the robots we spend the most time with.

It's odd to think of our devices as robots. But our phones, tablets, laptops, smartwatches, PCs, and connected home devices are, in fact, conduits for some of the most advanced forms of AI ever created. Companies like Facebook, Google, and Twitter have built sophisticated, planetary-scale machine-learning algorithms whose entire purpose is to generate engagement—which is to say, to short-circuit your brain's limbic system, divert your attention, and keep you clicking and scrolling for as long as possible.

These technologies have fundamentally changed what it means to use a device. Steve Jobs famously described the personal computer as "a bicycle for the mind," and for years, the metaphor fit. Like bicycles, computers could help us get places faster, and reduce the effort needed to move ideas and objects around the world. But these days, many of our devices (and the apps we install on them) are designed to function less like bicycles, and more like runaway trains. They lure us onboard, tempting us with the possibility of rewards—a new email, a Facebook like, a funny TikTok video. Then, once we're in, they speed off to their chosen destination, whether it's where we originally wanted to go or not.

That these forces are largely invisible doesn't make them any less real. The algorithms that power platforms like Facebook and YouTube are many times more powerful than the technology that sent humans to the moon, or even the technology that allowed us to decode the human genome. They're the products of billions of dollars of research and investment, exabytes of personal data, and the expertise of thousands of

Ph.D.s from the top universities in the world. These AIs represent the kind of futuristic superintelligence we saw in sci-fi movies as kids, and they stare out at us from our screens every day—observing us, adapting to our preferences, figuring out what sequence of stimuli will get us to watch one more video, share one most post, click on one more ad.

Humans have worried about the degrading psychological effects of machines for centuries. (Adam Smith warned in *The Wealth of Nations* that automated factory equipment was making us "as stupid and ignorant as it is possible for a human creature to become.") And in recent years, sounding the alarm about the negative consequences of smartphone use has become a thriving cottage industry. We now have "screen detox" resorts for adults, screen-time consultants for kids, and "digital sabbath" groups that encourage members to unplug completely for one day a week. We have even invented new phones to solve the problems with our old phones—like the Light Phone, a $250 "dumbphone" with a black-and-white display that can only be used for calling and texting.

Again, I'm not a screen-time fundamentalist, and my goal isn't to convince you that you're using your phone too much. (Even though you may well be.) Instead, I want you to do something I wish I'd done years ago: to take an honest, searching look at the relationship you have to your devices, and ask yourself the question, *Who's really in charge here?*

The answer to this question is important for a few reasons.

First, in order to do the kind of deeply human work we will need to do in the coming years—all of the social, surprising, and scarce tasks that will separate us from the machines—we need to be in control of our own bodies and minds, and be able to harness and direct our own attention.

Second, we need to understand the way that ceding control to our devices can harm our relationships with other humans. The psychologist Sherry Turkle has written at length about the phenomenon of "phubbing"—a strange-sounding but useful neologism that is short for "phone snubbing," and that describes the act of avoiding interactions with someone in favor of using your phone. She writes that phubbing amounts to "a flight from conversation—at least from conversation that is open-ended and spontaneous, conversation in which we play with ideas, in which we allow ourselves to be fully present and vulnerable."

Studies have shown that phubbing—or merely having our phones near us while we're interacting with other people—makes it harder for us to have enjoyable experiences with other people. One such study, conducted at the University of British Columbia, observed more than 300 people sharing a meal at a restaurant with friends and family. Half the participants were told to keep their phones on the table, with the ringer or vibrate setting turned on. The other half were told to silence their phones and put them away in a container. After the meal, the participants were given a questionnaire about their experience of the meal. People in the phones-on-the-table group reported enjoying their meals less, and being more bored and distracted, than people in the phones-in-a-container group.

All the evidence we have suggests that what matters is *how* we use our devices, not just how often we pick them up. Studies have suggested that certain types of device use are better for our mental well-being than others. For example, using Facebook passively (scrolling through our feeds, watching videos, absorbing news updates) has been shown to increase anxiety and decrease happiness, while using Facebook more actively (posting status updates, chatting with friends) has been shown to have more positive effects.

Which brings me to the third reason it's important to demote our devices, which is that by letting our smartphones and other devices steer our lives, we're missing out on many of the amazing, humanizing things we can use these machines to do.

For me, this hit home during the early days of the Covid-19 pandemic, when screen-based activities became my primary forms of social interaction. I attended happy hours and game nights over Zoom, held long FaceTime calls with my family across the country, and sent flurries of group texts with my closest friends.

What these positive experiences had in common was that they involved other people, and that while technology made them possible, I ultimately chose them, controlled them, and set the terms of my engagement. I didn't get manipulated into these interactions by a clever UX hack or an algorithm's invisible nudges. And while the companies whose tools powered those interactions may have benefited from my engagement, they gave me something of real human value in exchange for my attention and my data.

The difference between using our devices in a way that amplifies our humanity and a way that diminishes it, in other words, usually comes down to who's doing the driving.

· · ·

On the way home from the Alvin Ailey performance, I remembered a woman who had emailed me months earlier. Her name was Catherine Price. She was a science journalist who had written a book called *How to Break Up with Your Phone,* which outlined a thirty-day program she'd developed to help people like me recover from excessive phone use and form healthier relationships with their devices.

When I got home, I wrote her an email, and begged for her help. Mercifully, she said yes.

Before starting me on her phone detox plan, Catherine worked with me to figure out why I wanted to change my habits in the first place. She asked me to fill out an intake survey that asked questions like:

Why do you want to "break up" with your phone? What are you hoping to get out of the experience?

What do you like about your phone / want to continue doing?

What do you not love about your phone / want to spend less time doing?

In my responses, I poured my heart out. I told Catherine about my feelings of machine drift, and my fears that I was becoming blander and more predictable as a result of all my technology use. I confessed that I had found myself losing interest in interactions that didn't offer the same easy dopamine payout of a Twitter pile-on or a contentious Facebook thread, and that conversations with my friends and loved ones—

amazing and kind humans whose values and opinions I cherish—had started to feel less gratifying than getting online affirmation from strangers. I said that I didn't want to give up my devices entirely—I needed them for my job—but that I wanted to de-center them in my life and rebuild some of the willpower and self-control I'd lost. And I showed her my screen time statistics—which revealed that I usually spent between 5 and 6 hours a day on my phone and picked it up between 100 and 150 times.

"That is frankly insane and makes me want to die," I wrote to her.

"I will admit that those numbers are a bit horrifying," she replied.

Catherine's first piece of advice was to put a rubber band around my phone.

The rubber band, she explained, served two purposes. First, it was a tiny, physical speed bump for my fingers. It wouldn't keep me from using my phone—I could still tweet and text to my heart's content—but it would create an extra bit of friction. Second, the rubber band would serve as a constant reminder to be mindful. Every time I saw it, I'd notice that I was reaching for my phone, and I'd be able to stop myself to ask whether I really needed to check it, or whether I was just killing time.

The point of her phone detox plan, Catherine explained, wasn't getting me off my phone completely. It was about identifying the root causes of my phone addiction, including the emotional triggers—in my case, primarily boredom and anxiety—that caused me to reach for my phone in the first place. Then I could find other ways to satisfy those urges.

Ultimately, she said, the goal wasn't total abstinence. It was mindfulness.

"Your life is what you pay attention to," she told me. "If you want to spend it on video games or Twitter, that's your business. But it should be a conscious choice."

With a rubber band wrapped around my phone, I began noticing many of the strange subconscious rituals I'd developed. I noticed that I reached for my phone every time I stepped onto the elevator at my office, every time I left the front door of my apartment, and—most bizarrely—every time I inserted my credit card into a reader at a store, to fill the three-second gap before the card gets accepted.

I also realized how dependent I'd become on my phone as a source of constant, uninterrupted stimulation. I'd gotten used to walking around with AirPods on, listening to music and making phone calls. I watched YouTube videos while folding laundry, and Netflix shows while cooking dinner. I even wore waterproof headphones in the shower, so that I could listen to podcasts while I shampooed my hair.

I told Catherine about this. She laughed and said that I'd correctly diagnosed my problem.

"It's not really the phone," she said. "The phone is just the drug delivery device. The bigger issue is figuring out how to be alone with your own mind."

Psychologists call this problem "idleness aversion." Research has shown that being alone with our thoughts makes many of us profoundly uncomfortable, and that we generally even prefer physical pain to quiet solitude. In one experiment, conducted at the University of Virginia, college students were asked to sit alone in an empty room for a "thinking period" of ten to twenty minutes. They were also hooked up to electrodes, and told that if they wanted to, they could push a button to give themselves a painful electric shock. (They were under no obligation to give themselves the shock, and the experiment wouldn't end sooner if they did—it was purely an optional way they could distract themselves from their boredom.)

When the researchers looked at the results, they found that 71 percent of the men in the study and 26 percent of the women had shocked themselves at least once. The majority of the participants, confronted with the choice between sitting still and electrocuting themselves, had opted for the shock.

"The untutored mind," the researchers concluded, "does not like to be alone with itself."

If I was going to demote my phone, I needed to conquer my own idleness aversion. So, for days, I practiced doing nothing. During my morning walk to the office, I looked up at the buildings around me, and kept my phone firmly in my pocket. On the subway, I people-watched instead of listening to a podcast or tapping out emails. When a friend ran late for our lunch, I sat still and stared out the window.

I told Catherine how hard this was for me, and how often I'd been tempted to reach for my phone in search of stimulation. She said that was natural, and she reminded me that the point of the detox program wasn't just to use my phone less, but to rediscover the things I found fascinating and energizing about the offline world.

"Think of the bigger picture of what you're getting by not being on Twitter all the time," she said.

The other alarming thing I told Catherine was that, because I wasn't using my phone to fill downtime, I was noticing how many *other* people used their phones to cope with their own idleness aversion. Everywhere I looked, I saw a sea of bowed heads, peering into glowing screens, and it legitimately scared me.

Catherine said that was a common experience for her clients.

"I compare it to seeing a family member naked," she said. "Once you look around the elevator and see the zombies checking their phones, you can't unsee it."

After a few days with a rubber band on my phone, I started following Catherine's other recommendations. I kept my phone outside my bedroom at night so that it wouldn't interrupt my sleep. I pruned my apps, deleting distracting time-wasters and moving more calming, productive apps to my home screen, and disabled all but the most urgent push notifications.

Then, I started rebuilding my attention span by reading books—setting a timer and sitting down for ten minutes at a time, then twenty, then a full hour. I took daily walks without my phone, and I picked up hobbies, like cooking and pottery making, that kept my hands busy and my mind distracted from what was happening on Twitter.

Eventually, I started to acclimate to the feeling of being unstimulated, and I found strange things happening. The physical world seemed brighter and more alive. During my phone-free walks, I noticed little details I'd never noticed before—the misspelled "chicken parmesean" sign on the Italian restaurant down the block, the stately maple tree on the corner. My sleep and my mood improved, and I daydreamed for the first time in years.

The final phase in Catherine's plan is the "trial separation"—a twenty-four-hour period in which you don't use your phone at all. (I'm an overachiever, so I aimed for forty-eight hours.) I booked an Airbnb on a farm a few hours away, set my out-of-office autoresponder, and my wife and I took off for a weekend of off-the-grid leisure.

A phone-free mini-vacation involved some complications. Without Google Maps, we got lost and had to pull over for directions. Without Yelp, we had trouble finding open restaurants. But mostly, it was an amazing two days, filled with the kinds of small, subtle pleasures I hadn't experienced in years. I woke up at dawn, brewed strong coffee, and went for long hikes. We read books, did the crossword puzzle, and fell asleep to the sound of a crackling fire. I felt like a nineteenth-century homesteader, if the homesteader periodically worried that he was missing some good TikToks.

In the end, Catherine's thirty-day phone detox plan did reduce my screen time. My average daily phone use plummeted from nearly six hours a day to just over an hour, and I picked up my phone only about twenty times a day, roughly 80 percent less than I had at the beginning.

But it produced some other, harder-to-measure benefits, too.

First, demoting my devices made me more appreciative of

the technology in my life. For years, I'd viewed my phones and laptops as albatrosses, the burdens I had to bear as the price of a modern existence. But after a month of quasi-separation, I started seeing them with wonder and amazement, the way I had as a first-time BlackBerry owner. I marveled at the way I could, with a few taps and swipes, call up any piece of information ever recorded, or talk with almost anyone in the world. My texts and emails felt less like boring upkeep, and more like joyful interactions. The internet began to feel more like an earlier, healthier version of itself.

Second, demoting my devices made me much more productive—not in the strict economic sense of performing more labor, but in the literal, generative sense of *producing* more: more new ideas, more inspired solutions to problems, more compelling interactions with other people. When I took the cognitive and creative energy I'd been sinking into maintaining my digital presence and applied it elsewhere, I found myself coming up with all kinds of projects I wanted to pursue (including writing the proposal for this book). And because I wasn't amped up on phone-supplied adrenaline and cortisol all day, I actually had the energy to pursue them. I also felt more emotionally perceptive, and had an easier time picking up on other people's moods—noticing the subtle nonverbal cues I would have missed if I'd been staring at Twitter.

The third big result of demoting my devices, and the one I least expected, was the effect it had on everyone around me. I tried not to make a big deal of my phone detox while I was doing it—there is no more annoying human than one who is performatively into trendy wellness regimens—but my rubber-band-covered phone invariably came up everywhere: the office, coffee shops, airplanes. I found myself telling dozens of strangers about Catherine's plan (and probably sold a few hundred copies of her book). And I noticed that my improved focus resulted in other people being more mindful of their surroundings, too. During work meetings, my colleagues would notice me sitting still and listening intently, and they'd put away their own phones. At the park, other dog owners would see me taking enjoyment in watching my dog zoom around the meadow, and they'd look up and watch their own dogs more closely.

Jenny Odell, the author of *How to Do Nothing,* writes about how she overcame her own idleness aversion by getting really into bird-watching. And as she started noticing more birds in the air around her, she noticed that many of her friends started pointing out birds, too.

"One thing I have learned about attention is that certain forms of it are contagious," she writes. "When you spend enough time with someone who pays close attention to something (if you were hanging out with me, it would be birds), you inevitably start to pay attention to some of the same things."

I don't mean to imply that demoting your devices will cure all your worldly ills or turn you into an enlightened guru who wanders the earth dispensing rubber bands and warning about screen addiction. But in my case, reclaiming my attention and repairing my relationship to my phone did produce real, tangible benefits. And, looking back on my thirty-day detox, it was a prerequisite for any other steps I took to prepare myself for the future.

After all, smartphones and social media apps have real benefits, but they are also fundamentally extractive tools that exploit our cognitive weaknesses to get us to click on more posts, scroll through more videos, and view more targeted ads. They do this with the help of AI, which allows them to more accurately predict our preferences, steer our attention, and activate our brain's pleasure centers with flashy and exciting rewards. And by making it possible for us to be constantly stimulated, they deprive us of the opportunity to be bored, to let our minds wander, to cross-pollinate ideas and get lost in our imaginations—experiences that are central to our humanity, and without which we might as well be robots.

A few weeks later, I published a column about my phone detox in the *Times,* and I found out just how unoriginal my experience was. Millions of people read the story, and it became, by an order of magnitude, the most popular thing I've ever written. Catherine and I got invited on the *Today* show to teach Kathie Lee Gifford and Hoda Kotb how to detach themselves from their phones, and I got hundreds of emails and comments from readers about their own phone addiction sagas.

One reader wrote: "Thank you for your latest article. While reading the article, I realized that I too am addicted to my cell phone. For God's sake, I check it right before I open the door to leave the house, five seconds later after descending the staircase and locking the front door, then on the six seconds to my car, then right when I get in the car. I can only imagine how many times I check it throughout the day."

Another reader wrote: "I speak to children and parents about online safety in Ireland. Everyone wants to be told what software they can install to monitor their kids' online usage for them, but the key problem, as you've pointed out, isn't really what they're doing online, but what they're not doing offline. The inability to read long books/watch movies/concentrate on ANYTHING for longer than 30 seconds worries me greatly and is something that I feel more parents should also be concerned about."

But the best piece of feedback I got, by far, was from my wife. She'd seen my phone use spiral out of control for years and had begrudgingly accepted it as part of my base personality. But after I started my phone detox, she noticed that my behavior at home was changing. We'd started going to movies, taking pottery lessons together, and having long, meaningful conversations. I was asking more questions, paying better attention, and laughing more than usual.

One night, as we were watching TV on the couch together, she looked around for my phone, as if expecting me to grab it during a lull in the show.

I told her I'd put it away for the night, and she smiled.

"I feel like I've gotten you back," she said.

Rule 4

Leave Handprints

Men were not intended to work with the accuracy of tools, to be precise and perfect in all their actions. If you will have that precision out of them, and make their fingers measure degrees like cog-wheels, and their arms strike curves like compasses, you must unhumanize them.

—JOHN RUSKIN

Mitsuru Kawai should have been panicking.

It was 1966, and Kawai was an eighteen-year-old junior worker at the Toyota plant in Aichi Prefecture, Japan. He had spent the past three years studying at the Toyota Technical Skills Academy, apprenticing under the *Kami-sama,* or "gods"— Toyota's term for the master craftsmen who knew how to make every part of a car by hand. Kawai wanted to be like the *Kami-sama* someday, and he'd sought out a job in the local Toyota plant's forging division, where he could spend every day practicing taking glowing-hot rods out of a furnace, placing them on his anvil, and carefully hammering them into crankshafts.

Auto manufacturing was a solid, middle-class job, but it was increasingly looking like an endangered one. A few years earlier, General Motors—then the biggest automaker in the world—had installed the world's first industrial robot: a four-thousand-pound, one-armed behemoth known as the Unimate. The Unimate became a pop culture sensation, even appearing on *The Tonight Show* with Johnny Carson, where it wowed the audience by showing off various tricks it had been programmed to do, including putting a golf ball, pouring itself a glass of beer, and conducting the *Tonight Show* band. And it caught the eye of auto executives around the world, including at Toyota, who saw its potential to speed up their production and lower their costs. (What the robot *didn't* do was also presumably appealing; one TV commercial for the Unimate bragged that "he never complains, asks for a promotion, or demands a pay raise.")

Faced with these new, powerful robots, Kawai and his colleagues on the Toyota shop floor had some hard choices to make. Lots of experts and union leaders were predicting a bleak future for autoworkers, and blue-collar factory workers of all kinds. A 1961 article in *Time* magazine predicted the rise of "The Automation Jobless." Another article called factory automation "a ghost which frightens every worker in every plant."

But Kawai wasn't haunted by ghosts, and he didn't panic, or start looking for jobs in less threatened industries. Instead, according to a profile in the *Japan Times*, he decided to improve what, in Japanese, is called *monozukuri*, which translates as "making things," although its closest English equivalent is probably "craftsmanship." At Toyota, *monozukuri* referred to all the skilled, specialized labor that went into auto production. By focusing on becoming the best craftsman he could be, Kawai hoped that he would still be able to add value even if robots learned to do his specific job.

Over the next several years, Kawai developed a sixth sense for the tiny, subtle details of car production. He could look at a malfunctioning machine and tell, just from the sounds and smells it was emitting, what was wrong with it. He could estimate the temperature of molten iron, just by seeing which shade of red it glowed. And he was especially good at spotting the advantages humans had over robots. He noticed one day, for example, that there was a flaw in the undercarriage of one of Toyota's vehicles, stemming from a technical limitation in one of the robots it used to weld large metal pieces together. Kawai knew that an experienced human welder would do a cleaner job. So he convinced his bosses to de-automate that process, putting it back into human hands, which led to sturdier undercarriages and happier customers.

At a time when many autoworkers were either trying to outwork their new robot competitors or rebelling against them outright, Kawai's obsessive focus on *monozukuri* made him something of an outlier. He wasn't anti-automation, exactly. But he believed that factory robots were worthless without skilled humans around to teach them, to work alongside them, and to catch their mistakes before they cascaded into expensive failures.

"We cannot simply depend on the machines that only repeat the same task over and over again," he told a reporter. "To be the master of the machine, you have to have the knowledge and the skills to teach the machine."

In the late twentieth and early twenty-first centuries, as Toyota's factories became even more automated, Kawai's decision to focus on his human skills kept paying off. He got promoted over and over again, and his philosophy of *monozukuri* became a kind of rallying cry for Toyota's blue-collar workforce, who took pride in their work and wanted to avoid becoming mere robot handlers. Skilled craftspeople also often produce better cars with less waste, and in recent years, under Kawai's leadership, Toyota has bucked the automation trend by de-automating many of its production lines, bringing in humans to do jobs that were once performed by robots.

Today, Kawai is a living legend at Toyota. Workers call him "Oyaji," or father. He's the only employee in the company's eight-decade history to work his way up from the company's training academy to the executive ranks, and one of the only senior officials without a college degree. In 2020, Toyota named Kawai its first-ever chief *monozukuri* officer. The title reflected his decades of devotion to Toyota's workers, and his unyielding belief that even in the age of advanced robotics, their human skills made all the difference.

The situation Mitsuru Kawai faced as an eighteen-year-old autoworker in 1966 is roughly the same as the one that millions of educated white-collar knowledge workers find themselves confronting today. We worry, for good reason, that robots are on the verge of taking over our jobs and making us obsolete. And we're looking for something that can give us an enduring edge.

One option, of course, is to try to differentiate ourselves through hard work. This strategy has become increasingly popular in recent years, with the advent of so-called "hustle culture." All over social media, influencers and business gurus preach the value of productivity and constant, ceaseless effort. They post inspirational "hustle porn" memes on Twitter, LinkedIn, and Instagram with phrases like "Rise and Grind" or "Thank God It's Monday." They trade life hacking tips and cut out unnecessary cognitive burdens by wearing the same clothes every day or eating the same thing at every meal.

Hustle culture has a long lineage. In the late eighteenth and early nineteenth centuries, a former steelworker named Frederick Winslow Taylor came up with a theory of "scientific management" that took the American business community by storm. Taylor believed that most jobs could be broken into standardized, measurable tasks, and that those tasks could be perfected over time by ironing out inefficiencies and shaving away every millisecond of wasted time. Ultimately, he believed that enhanced productivity would be a win-win: companies would increase their output, and workers would get the satisfaction of operating at peak performance.

Today's version of Frederick Winslow Taylor is probably Gary Vaynerchuk, a marketing guru and social media influencer who has made a lucrative career out of inspiring his millions of followers to hustle harder. ("You need to work every goddamn minute you can," Vaynerchuk said in a 2018 YouTube video.) But there are plenty of contenders. Elon Musk, the Tesla and SpaceX founder, famously works himself to the point of burnout, even sleeping on Tesla's factory floor during intense production cycles. ("Nobody ever changed the world on 40 hours a week," Musk once tweeted.) Marissa Mayer, the former chief executive at Yahoo, bragged in a 2016 interview about how hard she worked, saying that it was technically possible to work as many as 130 hours a week "if you're strategic about when you sleep, when you shower, and how often you go to the bathroom."

Unlike Taylor's scientific management, which was often mandated from the top down, hustle culture is typically self-imposed. It's an outgrowth of the philosophy the writer Derek Thompson has called "workism"—the belief, common especially among type-A millennial overachievers, that work is not just an economic necessity but the primary source of identity and meaning in our lives.

There are plenty of reasons to reject hustle culture. It carries real risks to workers' physical and mental health. It tends to favor young, childless, able-bodied men, who are less likely to have family responsibilities and more likely to be able to work punishingly long hours. And it reinforces a brutal, regressive capitalist ethos that can undermine efforts to make workplaces more equitable and humane.

But I want to draw your attention to a more immediate problem with hustle culture, which is that in the age of AI and automation, hustling is actually counterproductive. No matter how hard you work, you simply cannot outwork an algorithm. If you try, not only will you lose, but you will sacrifice your unique human advantages in the process.

The idea that we can outwork machines is a seductive fantasy, going all the way back to the legend of John Henry and the steam engine. But many of today's most powerful technologies operate at such a vast scale, with such enormous computing power behind them, that the idea of competing with them head-on isn't even conceptually possible. What would it even mean for a human librarian to "compete" with Google at retrieving information from among billions of websites? Or for a human stock trader to "compete" with a high-frequency trading algorithm that can analyze millions of transactions per second? More to the point, why would they even want to try?

Instead of trying to hustle our way to safety, we should do what Mitsuru Kawai did—refusing to compete on the machines' terms, and focusing instead on leaving our own, distinctively human mark on the things we're creating. No matter what our job is, or how many hours a week we work, we can practice our own version of *monozukuri,* knowing that what will make us stand out is not how hard we labor, but how much of ourselves shows up in the final product.

In other words, elbow grease is out. Handprints are in.

A few years ago, Yann LeCun, then the head of Facebook's AI research division, articulated his own theory of the value of handprints.

LeCun is one of the so-called "godfathers of deep learning," and one of a handful of scientists who pioneered the use of deep neural networks, an AI technique that now powers much of the consumer internet. He was giving a lecture at an

MIT conference, and for nearly an hour, he spoke about all the dense, technical topics you'd expect: adversarial training, scalar rewards, multilayered convolutional networks.

Near the end of the talk, LeCun made an unexpected prediction about the effects all of this AI and machine learning technology would have on the job market. Despite being a technologist himself, he said that the people with the best chances of coming out ahead in the economy of the future were not programmers and data scientists, but artists and artisans.

To illustrate his point, he projected a slide with two photos: one of a Blu-Ray DVD player, which was being sold on Amazon for $47, and another of a handmade ceramic bowl, which was selling for $750. The difference in complexity between the two objects, he said, was stark. The Blu-Ray player was a sophisticated piece of technology with hundreds of parts, assembled by robots in a cutting-edge factory, whereas the bowl was a simple object made out of clay on a wheel, using techniques that are thousands of years old. And yet, the bowl was selling for nearly twenty times the price.

"There is real human intervention, real human experience" in the bowl, he explained to the crowd. And in the future, he predicted, "we're going to give more value to those, and less and less value to material goods that are built by robots."

I've heard versions of LeCun's prediction from a number of leading AI experts and economists. And despite my sometimes skeptical view of expert predictions on these topics, I've come to believe in it.

Why? Because, well, look around. Our economy is awash in mass-produced consumer goods that used to function as status symbols—flat-screen TVs, dishwashers, hot tubs—but have now become cheaper and more widely accessible. These days, a better indicator of luxury is how *little* technology is involved in producing the things you consume. Handmade furniture, bespoke clothing, custom art on your walls—these are the kinds of purchases that signal high status, precisely because they require a lot of human work.

Social scientists call this the "effort heuristic," and it's one of the better-documented phenomena in consumer psychology. In his book *The Power of Human,* Adam Waytz, a psychologist and professor at Northwestern University's Kellogg School of Management, runs down a laundry list of studies showing that people greatly prefer goods and experiences that have obvious human effort behind them, even when the goods and experiences are identical to those produced by machines.

One such experiment, led by Kurt Gray, a psychology professor at the University of North Carolina, gave identical bags of candy to two groups of participants. All of the candy was randomly selected, but one group was told that their candy had been personally picked out for them by another person. That group rated their candy as better tasting than the group that was told that their candy had been randomly selected. In another of Gray's experiments, participants who were given massages by electronic massage chairs reported experiencing more pleasure when they were told that a human had pushed the button to activate the chair.

The effort heuristic explains a lot about the rise of craft breweries, farm-to-table restaurants, and artisanal Etsy shops. It explains why vinyl records and printed books are still popular, even as streaming music services and e-books have become widely available, and why high-end cafés can still charge $7 for a cappuccino, even though most of us have machines capable of making perfectly good coffee in our homes and offices.

It also explains why the inverse is true—that when we hide or eliminate the human effort behind something, we often devalue it. My favorite example of this is Facebook birthdays. In the early days of Facebook, getting congratulatory messages on your birthday was a genuinely special event. It meant that your friends were thinking of you, and that they'd cared enough to spot your birthday, go to your profile, and think of something nice to write on your Facebook wall. But as years went on, Facebook tried to spur more engagement by making birthday greetings as easy and frictionless as possible. It allowed users to export their friends' birthdays to their calendar apps, placed them in a prominent spot on the news feed, and even began auto-populating stock birthday messages you could post with a single click.

As a result, Facebook birthday messages not only lost their special, intimate value—they actually became inversely related to intimacy. You knew that everyone who wrote "Happy birthday!" on your Facebook profile had merely done what the app told them to do, and that they didn't care enough to send you a more personalized message. By reducing the effort involved in wishing someone a happy birthday, Facebook had turned a caring expression into a mild insult.

Internalizing the handprints principle—which says that the more obvious the human effort behind something, the higher its perceived value—is a crucial part of preparing for the future.

Because, let's face it: we know that AI and automation are going to make a lot of things *very* easy. Handling packages, creating sales projections, driving from Point A to Point B—these tasks, and thousands more, will go from requiring lots of human labor to requiring little or none. And the humans who currently get paid to do those tasks will need to get creative and figure out ways to make the value of their contributions more evident.

I note, carefully, that leaving handprints is not just about showing off, or taking credit for as much work as possible. It's also different from hustle culture, which is all about performative productivity. Hustling is about how hard we work; leaving handprints is about how *humanely* we work.

Often, leaving handprints is a simple matter of making invisible labor visible. For designers, it might mean walking your clients through your creative process step-by-step, so they understand how much labor and expertise go into each sketch. For software engineers, it might mean practicing explaining what you're doing in plain English to nontechnical executives.

Other times, it's a matter of putting in effort that is strictly unnecessary, but greatly appreciated. For insurance agents, it might mean sending sympathy cards to clients who have just lost a home in a fire or been involved in a car accident. For retail workers, it might mean getting to know your store's regular customers and setting aside special items you think they might like for the next time they come into shop.

For me, leaving handprints means that I start every reporting assignment by figuring out how I can put my unique stamp on it, and not have it feel like a generic story that any other re-porter (or any piece of AI software) could have written. That might mean punching up a dry technical explainer with jokes, or writing in the first person about my own experiences, rather than adopting the depersonalized, "view from nowhere" posture.

I also pay particular attention to certain high-impact ges-tures where a little humanity can go a long way. When my pub-lishers send out review copies of a new book, I make sure I write a handwritten note to every recipient, thanking them for taking the time to read it. When I finish a big group project, I often send homemade cookies or a small gift to thank my colleagues who helped me on it. When I write my yearly self-evaluation, I try to infuse it with voice and personality, so my bosses know I'm a real person and not an over-polished automaton.

None of this is new or revolutionary. In fact, it's kind of sad that I have to consciously remind myself to act like a human. But it does take conscious effort, especially when so much of the technology in our lives is designed to make things easy. Like many people, I've spent years absorbing the perfectionist tendencies of hustle culture, which teaches us that the key to success lies in optimizing our performance as if we were a sports car or a speedboat. And I've had to remind myself that in an era of labor-erasing technology, the most meaningful ges-ture is often the messy, imperfect one that sends the message: *Hey, I tried.*

Companies, too, will have to figure out a strategy for leaving handprints.

In their book *The Experience Economy,* business school

professors B. Joseph Pine II and James H. Gilmore write about the way certain enterprises move through the "progression of economic value." They start out selling commodities, eventually begin selling goods, morph into providing services, and ultimately end up designing experiences.

"Those businesses that relegate themselves to the diminishing world of goods and services will be rendered irrelevant," they write. "To avoid this fate, you must learn to stage a rich, compelling experience."

To illustrate their point, they use the example of coffee, which can be bought very cheaply from a wholesaler, slightly less cheaply from a supermarket, for $4 or so at Starbucks, or for $10 at a high-end café in Italy. In each transaction, money is being exchanged for coffee. But what customers are actually paying for varies. At a wholesaler, they're paying for the beans. At a supermarket, they're paying for the beans and the packaging. At Starbucks, they're paying for the beans, the packaging, and the service. And at a high-end café in Italy, they're paying for the beans, the packaging, the service, and the experience of having a cup of coffee in Italy, surrounded by Italians, maybe served by a charming barista who can explain the subtle flavors in their *caffè macchiato*.

Automation doesn't eliminate the need for coffee wholesalers or supermarkets or Starbucks. But it makes those kinds of businesses more fragile, and easier to compete with. And it gives the competitive edge to the businesses that sell experiences, which can't be copied or programmed into a machine nearly as easily.

Several years ago, Best Buy learned this lesson out of necessity.

Like lots of other big-box stores, Best Buy was struggling to keep up with Amazon and other online retailers. Sales of big-ticket items like TVs were falling, and many of the products that once lured customers into stores, like new-release CDs and DVDs, were becoming obsolete. When customers did come into stores, they were increasingly "showrooming"—that is, looking at an item, then going online to buy it for a lower price elsewhere. Its stock price had fallen off a cliff, it had been forced to close stores and lay off workers, and investors smelled death.

Then, in 2012, a new CEO, Hubert Joly, took over. Joly, a Frenchman with a lilting accent, was a former management consultant who believed that the only way Best Buy would survive was to stop trying to compete on Amazon's terms and start looking for another advantage. He realized, as he told me in a 2017 interview, that "if buying a technology product was purely a commodity game, we wouldn't have a chance."

So, he and his team came up with a strategy that involved turning Best Buy into a high-touch, face-to-face business that provided deeply human experiences that e-commerce retailers— and their sleek, hyper-optimized, robot-filled warehouses— couldn't match. The company invested in more training for store associates, and started an In-Home Advisor program that allowed customers to get personal consultations from trained Best Buy experts, who would come to their house to help them choose the right big-screen TV for their living room, or figure out which stereo system would sound best on their patio. The program, which launched in 2017, was an immediate hit, and created a core group of hyper-loyal customers who began treating Best Buy as a kind of personal tech concierge, rather than just a big-box store.

"The business we're in is not simply selling products—it's connecting human needs with technology solutions," Joly told me. "So, our focus is on these human needs."

Joly's humanist strategy brought Best Buy back to life. Sales skyrocketed, customers stopped showrooming, and within a few years, the company's stock price was at an all-time high, with happy workers and satisfied shareholders. Joly, who stepped down as Best Buy's CEO in 2019, was given a hero's farewell.

A much smaller example of a successful handprints strategy is what happened to Heath Ceramics, an iconic seventy-year-old pottery studio in Sausalito, California.

In 2003, when Heath's current owners, Catherine Bailey and Robin Petravic, acquired Heath from its original owner, it was a bit of a mess. Its ceramics had developed a cult following among Californian aesthetes—among its clients were the architect Frank Lloyd Wright and Alice Waters, the founder of Chez Panisse—but it was a creaky legacy business that was losing money, at a time when many small ceramics studios were going out of business, and cheap, mass-produced ceramics made in overseas factories were dominating the industry.

Bailey and Petravic realized, correctly, that they couldn't compete with foreign producers on price or volume. But they *could* compete on humanity. So, they embarked on a bold turnaround plan, which included many steps that no traditional cost-cutting consultant would have recommended. They turned down investment offers and kept Heath's manufacturing local, rather than sending it abroad. They moved their main production facility to a more expensive building in San Francisco's Mission District, and began offering factory tours, so that customers could see their bowls and mugs and tiles being made. They started the Heath Clay Studio, a kind of experimental lab where master potters could produce one-of-a-kind designs, and the Heath Newsstand, which sells hundreds of hard-to-find magazines from around the world. They also opened Heath Collaborative, a kind of artisanal maker bazaar adjacent to their San Francisco studio where jewelers, fiber artists, and bakers could set up shop.

As they humanized their facilities, Bailey and Petravic also experimented with putting humanizing touches on Heath's ceramics themselves, such as personalized stamps that told you exactly which of the company's employees had glazed each particular vase. Their goal in doing all these things was to show their work, and to remind customers that real humans—not machines—were creating these objects. And this strategy appears to have worked. Since 2003, Heath has mounted an impressive comeback. It has grown to more than two hundred employees and $30 million in annual sales, and last year, it became debt-free for the first time since the 2003 acquisition.

Like Hubert Joly at Best Buy, Bailey and Petravic weren't the most tech-savvy entrepreneurs on the planet. They didn't have cutting-edge AI at their disposal, or an army of well-paid programmers and logistics experts who could spend their time shaving milliseconds off every process. But they correctly sized up the competition and saw how they could stand out by emphasizing their own humanity, rather than erasing it. They succeeded by figuring out what Mitsuru Kawai figured out in a Toyota factory nearly sixty years earlier—that, when machines are our biggest competition, our humanity is worth more than our hustle.

Rule 5

Don't Be an Endpoint

*Right now, I feel like they need us, but they will find a
way to get rid of us if there's something that costs less.*
— VERLISA LEONARD, call center worker

On May 8, 2018, Google CEO Sundar Pichai took the stage at
a conference and did a mind-blowing demonstration of the
latest product to emerge from the company's vaunted AI research division.

The product was a voice-based AI assistant called Duplex,
which was capable of booking appointments, making restaurant reservations, and doing other tasks over the phone. To
demonstrate, Pichai played a recording of a call Duplex had
recently made to a hair salon.

"Hello, how can I help you?" the receptionist asked.

Duplex responded: "Hi, I'm calling to book a woman's haircut for a client. I'm looking for something on May third."

"What time are you looking for?" the receptionist asked.

"Twelve P.M.," Duplex said.

"We do not have a twelve P.M. available. The closest we have to that is a one-fifteen," the receptionist said.

"Do you have anything between ten A.M. and . . . uh . . . twelve P.M.?" Duplex asked.

"Depending on what service she'd like," the receptionist said. "What service is she looking for?"

"Just a woman's haircut for now," Duplex responded.

From there, the conversation proceeded in totally normal fashion. Duplex didn't stumble or get confused, and it even inserted some "hmms" and "uhs" for extra realism. At no point did the human on the other end of the conversation realize that she was talking to an AI.

When the call was over, the audience broke out in applause. I was watching the demo, slack-jawed, from home. And as I saw the commentary from amazed and terrified nerds flooding in, one particular reaction caught my eye. It was from Chris Messina, a former Google designer. He tweeted:

"Google Duplex is the most incredible, terrifying thing out of #IO18 so far . . . Example use case: Google Assistant calls a hair salon to book an appointment. The human booking the appointment has no idea she was talking to an AI. Humans quickly becoming expensive API endpoints."

I still think about Messina's tweet all the time—in particular, the last sentence: "Humans quickly becoming expensive API endpoints."

In programming, "endpoints" are special kinds of web addresses that allow programs to communicate with other programs through what's known as an API, or application program interface. If you're trying to move information from one app to another—say, if Tinder wants users to be able to upload their Instagram photos—Tinder has to write a piece of code that asks the photos endpoint of Instagram's API for permission.

Translated from coder-speak, what Messina was saying is that the human receptionist in the Google demo was serving as a point of connection between two pieces of software—Duplex and the salon's appointment calendar—and she was necessary only because those machines couldn't talk directly to each other yet.

That tweet ruined me, because after seeing it, I couldn't stop noticing human endpoints—people whose jobs mainly consisted of taking directions from a machine or serving as a bridge between two or more incompatible machines.

I'd see a security guard at an office building checking visitors into the building's security system and pressing the buttons to let them through the turnstile, and I'd think: *endpoint*.

I'd go to the doctor's office for my annual physical, and I'd see the nurse practitioner reading numbers off medical instruments and plugging them into an iPad loaded with my electronic health records, and I'd think: *endpoint*.

I'd see a Starbucks barista handing mobile delivery orders off to a Postmates courier—a human following one app's instructions and handing the product over to another person following a different app's instructions—and I'd think: *two endpoints*.

And, once in a while, if I'm being completely honest, I'd catch myself writing a story about something that happened on one website, posting it on another website, and promoting it on a third website, and I'd think: *yep, I'm kind of an endpoint, too*.

. . .

So far in this book, we've talked mainly about fully automated work—tasks that can be performed by machines from start to finish. But there is a lot of work that can only be partially automated, and we need to talk about those jobs, too.

Partially automated jobs come in two categories.

Category 1 consists of what you might call "machine-assisted" jobs. These are jobs in which humans direct and oversee the vast majority of the work and use machines as their helpers. An example of a machine-assisted worker would be a real estate agent who uses automated listing software to match clients with for-sale homes but shows up in person to take them to open houses and guide them through the home-buying process.

Machine-assisted jobs are the kind of jobs AI optimists generally have in mind when they talk about things like "human-centered automation." They're the jobs in which machines complement human workers, rather than displacing them.

Category 2 consists of what you might call "machine-managed" jobs. In these jobs, most of the work is directed and overseen by machines, and humans act as the gap-fillers, doing only the things the machines can't yet do on their own. Prominent examples of machine-managed jobs include gig work for companies like Uber, Lyft, and Postmates, along with the work performed by Amazon warehouse workers, Facebook and Twitter content moderators, and other people whose jobs consist mainly of carrying out instructions given to them by a machine.

Machine-managed jobs are less about collaborating with

AI systems, and more about serving them. An Uber driver is not "collaborating" with Uber's ride-matching algorithm, any more than a military cadet is "collaborating" with the drill sergeant who gives her marching orders. In these relationships, all of the power and leverage resides in the machines—humans are simply the plug-and-play accessories that follow orders, and that can be swapped out at will.

These machine-managed jobs are the endpoints, and they're a very dangerous place to be. Because often, the goal of a machine-managed job is simply bridging a technological gap that is in the process of being automated or training an automated system to achieve human-level performance.

In fact, it's exceedingly likely that every organization that uses machine-managed workers hopes to someday transfer those jobs to machines. Which means that machine-managed workers—the endpoints of the world—need to be on high alert. As the journalist Martin Ford writes in his book *Rise of the Robots*, "If you find yourself working with, or under the direction of, a smart software system, it's probably a pretty good bet that—whether you're aware of it or not—you are also training the software to ultimately replace you."

Many endpoint jobs exist in the industries you'd expect—service, retail, transportation. But they are also cropping up in more prestigious white-collar professions, where AI is turning what used to be machine-assisted work into machine-managed work.

In a 2018 *New Yorker* piece, "Why Doctors Hate Their Computers," Atul Gawande wrote that electronic medical software, which has become ubiquitous in U.S. hospitals over the last decade, was contributing to rising rates of burnout and depression among doctors by drowning them in record-keeping tasks and diverting them from patient interactions. That sentiment was echoed by Emily Silverman, a physician in San Francisco, who wrote in a 2019 *Times* op-ed that the electronic health records system used by her hospital had turned her and her colleagues into stressed-out machine minders.

"We are met with relentless reminders of tasks we haven't completed, supplications to correct our documentation for billers, and daily, jaundiced reminders," Silverman wrote.

The clash between doctors and their software perfectly illustrates the trade-offs of partial automation. In many ways, electronic medical records are a huge improvement on the status quo. When implemented properly, they improve patient safety, reduce costs, and limit the frequency of medical errors. But they have also made some doctors feel like their jobs consist exclusively of moving data from one screen to another. (One researcher called this feeling "technological somnambulism.") They used to be machine-assisted workers. Now? They're not so sure.

This feeling—a fear among workers of being downgraded by intelligent machinery, rather than assisted by it—isn't new. In fact, one of the best examples of it happened in 1970, when General Motors opened a brand-new, highly automated plant in Lordstown, Ohio.

Hailed in the press as the "plant of the future," the Lordstown factory was a dazzling tribute to modernity, with a fleet of twenty-six robots that zoomed around the factory floor. GM thought workers would enjoy spending their days in this futuristic paradise. But workers hated it. They were stressed out by the higher production quotas their managers set for them, and they felt dehumanized and bored by being made to operate machines all day. Here's how one Lordstown worker described his daily routine:

> You do it automatically, like a monkey or dog would do something by conditioning. You feel stagnant; everything is over and over and over. It seems like you're just going to work and your whole purpose in life is to do this operation, and you come home and you're so tired from working the hours, trying to keep up with the line, you feel you're not making any advancement whatsoever. This makes the average individual feel sort of like a vegetable.

In 1972, workers at the Lordstown plant got so fed up that they decided to strike. The strike drew national attention, and people across the country began talking about "Lordstown Syndrome"—a kind of existential malaise brought about by a new kind of partial automation. *Newsweek* called the Lordstown strike an "industrial Woodstock," and in 1972, a *New York Times* editorial called for GM to "be concerned with keeping alive the individual's sense of worth in the robot-ruled workplace."

After a twenty-two-day strike, GM caved. It scaled back production targets, gave workers more time off, and formed "humanization teams" tasked with improving conditions at the Lordstown plant. Edward Cole, then GM's president, said in a speech that the company had realized the error of making human workers subservient to machines.

"It is not machines but people on whom our future progress must depend," he said.

• • •

One group of people who have to be especially careful not to become endpoints are remote workers.

In a previous draft of this book, written before the Covid-19 pandemic, I had an entire chapter making the case that remote work was overrated. I cited research suggesting that our best, most human work happened when we saw other people face-to-face, rather than through our screens. In order to prepare ourselves for the AI-filled future, I argued, we needed to literally work *together*—in offices and on job sites where we could bump into co-workers, spark spontaneous discussions, and come up with creative ideas on the fly. For obvious reasons, I scrapped that chapter. (I enjoy a good hot take, but telling people to pack into crowded offices during a pandemic felt a few degrees too toasty.)

It's clear, now, that remote work is here to stay, and that virtual collaboration is going to be part of how many white-collar professionals operate. A July 2020 survey by Gartner found that 82 percent of corporate leaders plan to allow their employees to work remotely at least part-time after the pandemic, and nearly half of respondents said that they planned to allow full-time, indefinite remote work.

Shifting to remote work was clearly the right call during the pandemic, and it will continue to make sense for many workers with flexible remote work options to relocate away from high-cost cities. But I still believe—and the evidence still strongly suggests—that people who have regular, in-person contact with their colleagues have an advantage when it comes to doing the kinds of deeply human work we will need to do in the future.

I struggled to work well from home during the pandemic, and I'm not alone. As the pandemic dragged on, many workers found themselves frustrated by how hard it was to generate creative ideas, build team camaraderie, and onboard new employees over Zoom calls and Slack threads. They were also very, very tired from trying to juggle childcare and navigate virus precautions, and from staring at the same screens all day.

Executives were frustrated, too. Adobe CEO Shantanu Narayen complained that remote work was taking a toll on the company's ability to get new initiatives off the ground. "When you're trying to create a new project," Narayen said, "you want people around that water cooler." Reed Hastings, the chief executive of Netflix, called remote work a "pure negative" in an interview with *The Wall Street Journal*. Asked when he planned to bring his team back to the office, Hastings replied, "Twelve hours after a vaccine is approved."

It's no surprise that executives wanted to get back into offices. Studies have found that groups of people located in the same room solve problems more quickly than people who collaborate electronically, and that co-authors of academic papers who are located closer together geographically tend to produce higher-quality research. Studies have also shown that team cohesion suffers in remote work arrangements, and that while remote workers may be more productive than office workers, they're often less creative.

Remote work has plenty of practical benefits for workers, of course. Parents who work remotely get to spend more time with their kids, workers get to avoid stressful commutes, and people with disabilities can often work more easily in a home office. But these benefits come with trade-offs. Working from home blurs the line between work and leisure, and makes it harder for many people to disconnect and recharge. In addition, remote workers give up many of the social benefits of office culture, such as in-person bonding, mentorship, and career development. And they miss out on what John Sullivan, a management professor at San Francisco State University, calls "serendipitous interaction"—the random, chance encounters between workers at the cafeteria, or in line for the coffee machine, which often result in novel conversations and unexpected ideas.

But the biggest risk of remote work, when it comes to automation, is that it's much harder to display your humanity in the absence of face-to-face interaction. In some sense, remote workers are already halfway automated. They are experienced as two-dimensional heads in a Zoom chat, or avatars in a Slack thread. Their output is most often measured in terms of tasks completed and metrics hit, and their ability to contribute to an organization in subtler, more human ways—cheering up a demotivated co-worker, organizing happy hours, showing an intern the ropes—is dramatically limited.

Because of this, it's even more important for remote workers to go overboard in expressing their humanity and reminding others of their presence. And it's important for organizations that employ remote workers to bring those workers in for regular, in-person get-togethers so they can be fully socialized and integrated into their teams.

Even before the pandemic, companies were already ex-

perimenting with ways to make remote workers feel more socially connected. At GitLab, an open-source collaboration
platform, remote workers are encouraged to schedule "virtual
coffee breaks"—purely social video conferences—and join the
"random room," an always-on Google Hangout that functions
as a kind of virtual water cooler. At the Seattle-based software
company Seeq, one employee per day is responsible for "sharing time," a fifteen-minute video presentation about a non-
work topic of their choice. Automattic, the all-remote maker of
WordPress, organizes a once-a-year "grand meetup," a week-
long retreat during which the company's seven hundred–plus
employees work on collaborative projects by day, and go on
social outings at night.

These are good steps to make remote workers feel more
included. But they're still not a true substitute for the social
machinations of office life. So, if you're a full-time remote
worker, you should go out of your way to jam social interaction
into your days. Meet up with other remote workers at a restaurant near you. Form purely social group texts with your co-
workers. Start holding virtual potlucks or set up a rotating gift
exchange. Put yourself in positions where you can express your
humanity, and where there can be no mistaking you for a robot.

If you, like me, have a job that doesn't often require you to act
as an endpoint, consider yourself extremely lucky. And don't
rest on that luck. Stay aware of the developments in your field,
and any new technologies that might be used to change your
role to a more machine-managed one or move you further
away from the human parts of your job.

If you manage a team or lead an organization, make sure that the technology you're deploying is empowering people, rather than dehumanizing them. Invite workers into the decision-making process around automation and solicit feedback on how the machines are (or aren't) working for them. And remember that what happened in Lordstown—a high-tech, ultra-productive workplace brought to its knees by disillusioned workers who were fed up with being treated like robots—could happen to your organization, too.

If you're a machine-managed endpoint, or you think you might become one, you have a few choices to make.

First, if it's at all possible, *get out*. Endpoint jobs are dehumanizing and extremely vulnerable to automation, and because they're essentially stopgap measures until adequate technology arrives, they rarely improve over time. If your job consists mostly of moving information from one system to another, propose a different role for yourself that involves more complex, judgment-based work. If you're a digital ad buyer, ask to be involved earlier in your team's creative process. If you work in sales, propose sitting in on strategy meetings, rather than just building PowerPoint decks.

If you can't switch jobs, suggest changes to your existing job that would make your work more human, and give you more control over your tools. A number of unions have already successfully fought for these kinds of provisions for their members, and decades of labor activism in the manufacturing sector—like the Lordstown negotiations of the 1970s, which resulted in worker-led "humanization teams" being formed at GM—can give white-collar workers a model to follow.

If there is no possibility of changing your job and no way to avoid doing machine-managed work—either because you're a contract worker for a platform whose entire business model is based on machine-managed work, like Uber or Lyft, or because you don't have the leverage or authority to suggest changes to your job—then you need to come up with an escape plan.

One clear lesson from history is that people don't remain endpoints for long. There are simply too many incentives to finish automating these processes, and too many technologists working on taking humans out of the loop. When the day arrives that the machines can finally talk to each other, you don't want to be standing in the middle, wondering where your job went.

Rule 6

Treat AI Like a Chimp Army

We fixed a technical issue that caused incorrect transla-
tions from Burmese to English on Facebook. This should
not have happened and we are taking steps to ensure it
doesn't happen again.
    —a Facebook spokesman, apologizing for a flaw in the
company's machine learning AI that translated the name
of China's prime minister, Xi Jinping, as "Mr. Shithole"

Mike Fowler was up early when his phone started buzzing.

It was a Saturday morning in 2013, and Fowler—an American entrepreneur living in Melbourne, Australia—was used to
fielding business emails at odd hours. He was the founder of an
apparel company called Solid Gold Bomb, which sold novelty
T-shirts and other clothing items online, and there was always a
customer or employee somewhere who needed something.

But as he glanced down at his phone, he sensed that something was off. He saw Facebook messages from strangers, call-

ing him all kinds of awful names. An urgent email from a high-ranking Amazon executive. An interview request from the BBC.

"I was like, oh my God, something's happened," he told me.

His stomach dropped as he realized what had happened: his algorithm had betrayed him.

More than a year earlier, Fowler had hit upon an idea that had revolutionized his business. He had seen other T-shirt designers using targeted Facebook ads to sell shirts that combined popular phrases with messages aimed at hyper-specific audiences. (You may have noticed these shirts being advertised on your Facebook feeds in 2012 and 2013; they're the ones that said things like "Never Underestimate a CUBS FAN Born in AUGUST" or "Sorry, I'm Already Taken by a Sexy DENTAL HYGIENIST named TAMMY.")

Fowler wondered, What if you could use an algorithm to make these shirts not just for hundreds of different microaudiences, but millions? He wrote a simple script that took words from a dictionary and plugged them into popular catch phrases, automatically generated T-shirt designs out of them, and listed each one on Amazon in dozens of different colors and sizes.

Fowler launched his T-shirt algorithm on Black Friday of 2012, and by the following Monday he'd sold more than 700 shirts, more than he usually sold in a month. From there, it was off to the races. His algorithm generated 20 million different T-shirts online, using more than 1,100 different templates. Not all of these algorithmically generated T-shirts made sense, but it didn't matter. Creating Amazon listings was free, and the T-shirts were print-on-demand, meaning that they didn't need to exist in physical form until the right Cubs fan or dental hygienist came along to buy one.

It seemed like a brilliant plan until March 2, 2013, when a customer browsing Amazon noticed some disturbing shirts based on the popular slogan "Keep Calm and Carry On," whose messages had been generated using words Fowler had forgotten to exclude from the algorithm's word bank. They included messages like:

KEEP CALM AND HIT HER

KEEP CALM AND KNIFE HER

KEEP CALM AND RAPE A LOT

The offended customer posted images of the shirts on Twitter. An uproar followed.

When he grasped what was happening, Fowler went on Facebook and tried to explain that the messages had been algorithmically generated, that no human had ever seen or approved them, and that the T-shirts had never actually existed in physical form. But it was too late. Amazon barred Solid Gold Bomb from its store for violating its rules, and days later, Fowler was forced to lay off all of his employees and shut down the company.

A simple mistake—forgetting to remove potentially offensive words from a database of common verbs—changed the course of Fowler's life. I talked to him six years after the T-shirt incident, and he told me that he was still haunted by what had happened.

"It's been a tough slog," he said recently. "I've never fully recovered."

If an army of a thousand chimpanzees showed up at your office one day, looking for work, what would you do?

Realistically, you'd probably lock the door and call animal control, or tell yourself to lay off the magic mushrooms. But let's suspend reality for a second and imagine that, instead of panicking, you actually tried to find a task for them to do.

After all, under the right circumstances, chimps could make great workers. They're strong, agile, and fairly intelligent. They can be trained to recognize faces, pick up and carry items, and even respond to simple commands. You could imagine a group of well-trained office chimps loading and unloading warehouse shipments or restocking an empty laser printer.

Before you made any promises, of course, you'd want to know more about the chimps. How well-behaved were they? Did they have a history of aggression? How much training and supervision would they need? And ultimately, if you did decide to invite the chimp army into your office, you wouldn't do it right away. You might conduct a Chimp Safety Audit or convene a Chimp Oversight Task Force. You might decide to put a small number of chimps in a room under close supervision, train them to do a simple task, and evaluate the results before giving them more important assignments.

But whatever your risk tolerance was, I'm fairly confident that you wouldn't just invite the chimps in, give them badges and lanyards, and say "Okay, get to work!" And you sure as hell wouldn't *put them in charge*.

. . .

You probably see where I'm going with this.

In the last chapter, we discussed what happens when humans are turned into endpoints—when a process can't be fully automated yet, and people are called in to fill the gaps in the meantime. But the opposite problem also exists. Many organizations have made the mistake of *overautomating*—giving machines tasks and authority they really aren't equipped to handle and being surprised when things go horribly wrong.

Over the past several years, I've met lots of corporate executives who have been convinced that AI is an incredible, transformative technology, and that using AI in the workplace is a no-brainer decision, no more consequential or disruptive than switching the salad dressings in the cafeteria. They're itching to put as many machines to work as possible—not just boring RPA bots in the back office, but real AI, capable of making critical front-office decisions about strategy and operations. In fact, some of these executives are eager to give AI big pieces of their own jobs. In a recent article in the *MIT Sloan Management Review,* two executives predicted the rapid rise of "self-driving companies," businesses in which human managers are few and far between, and most corporate decisions, including hiring and firing, are made by algorithms.

Respectfully, these people are out of their minds. In fact, if you talk to the computer scientists working on the front lines of AI and machine learning, they'll tell you that even the best AI in existence is still far from being worthy of this kind of carelessly placed trust.

Today, most AI is similar to an army of chimps. It's smart, but not as smart as humans. It can follow directions if it has been properly trained and supervised, but it can be erratic and destructive if it hasn't. With years of training and development, AI can do superhuman things—like filtering spam out of a billion email inboxes or creating a million personalized music playlists—but it's not particularly good at being thrown into new, high-stakes situations.

One of my favorite websites is AI Weirdness, a blog written by a computer scientist named Janelle Shane, who started it while she was studying neural networks, a type of AI that mimics the way the human brain processes information, in a Ph.D. program at the University of California, San Diego. Shane noticed that sometimes, when she trained a neural network to do a task, it would fail in strange ways. Once, she trained a neural network to generate names for cats, using a data set of more than eight thousand cat names she got from an animal shelter. The program ingested the list of actual cat names and generated new, made up ones. They included:

Jenderina
Sonney
Mrow
Jexley
Pickle
Marper
Foppin
Toby Booch Snowpie
Big Wiggy Bool

Another time, Shane trained a neural network to generate cocktail recipes. The first result the neural network spit out was for a drink called "Morale and Phop Ngaba," whose recipe went as follows:

1½ oz lineappl
1 lunces crilpi juice
1 teaspoon sramge juices
Add witeasples
Fttr into a cocltail glass

Shane, who expanded AI Weirdness into a book called *You Look Like a Thing and I Love You,* would probably accuse me of giving AI too much credit by comparing it to chimpanzees. (In fact, she makes a different zoological comparison, writing that "AI has the approximate brainpower of a worm.") She writes that without close human oversight, AI is capable of making not just funny, harmless mistakes but also truly dangerous ones.

"Because AIs are so prone to unknowingly solving the wrong problem, breaking things, or taking unfortunate short-cuts," Shane writes, "we need people to make sure their 'brilliant solution' isn't a head-slapper."

But this cautionary note apparently hasn't made it to the decision makers of corporate America, because they still seem to be placing excessive trust in the wisdom of AI, often with consequences much more severe than accidentally advertising some offensive T-shirts or coming up with a disgusting cocktail recipe.

A trading firm called Knight Capital, for example, lost

$440 million in a forty-five-minute span on August 1, 2012, after an improperly installed automated trading system rapidly bought and sold millions of shares, pushing their prices up and creating massive losses when they had to be resold. The losses nearly put Knight Capital out of business, and the firm had to receive hundreds of millions of dollars in emergency financing in order to stay afloat.

Or take Watson, the IBM-owned AI that famously defeated a *Jeopardy!* champion in 2011. In 2013, IBM teamed up with The University of Texas MD Anderson Cancer Center to develop a new Watson-based oncology tool that could recommend treatments for cancer patients. But the program had flaws. In 2018, internal tests obtained by the health news publication *Stat* found that Watson had been improperly trained on data from hypothetical patients rather than from real ones and, as a result, made some faulty recommendations for treatment. In one case, Watson reportedly recommended that doctors give a sixty-five-year-old lung cancer patient with severe bleeding a type of medicine that could have worsened his bleeding. (IBM told *Stat* in a statement that it had "learned and improved Watson Health based on continuous feedback from clients, new scientific evidence, and new cancers and treatment alternatives.")

Flawed AI often disproportionately impacts marginalized people, because the data used to train the algorithms is often drawn from historical sources that reflect their own patterns of bias. Much of the arrest data used to train the "predictive policing" software used by law enforcement agencies, for example, reflects decades of systematic overpolicing of predominantly Black and Latino neighborhoods, as well as racially discriminatory policies like stop-and-frisk. One notorious law enforcement algorithm, known as Correctional Offender Management Profiling for Alternative Sanctions, or COM-PAS, has been used by courts to recommend sentences for criminal defendants, based on computer-generated predictions of how likely they are to re-offend. A 2016 investigation by ProPublica found that COMPAS was nearly twice as likely to label Black defendants as future criminals compared with white defendants.

All over the world, faulty and untested AI and automated systems are being entrusted with incredibly important decisions. And while some of the governments, companies, and organizations implementing these systems may be going about it the right way—thoughtfully and rigorously evaluating new algorithms, doing threat modeling and scenario planning to figure out what could go wrong, putting plenty of human supervision in place—many aren't. They're just throwing open the doors, letting the chimp army in, and praying for the best.

Yoshua Bengio, a computer scientist and pioneer in the field of deep learning, is one of the people you'd expect to be advocating for shoving AI into every major decision-making process. But in a 2018 interview with the journalist Martin Ford, he came out strongly against using AI to make important, life-changing decisions, such as how long a convicted felon should go to prison.

"People need to understand that current AI—and the AI that we can foresee in the reasonable future—does not, and will not, have a moral sense or moral understanding of what is right and what is wrong," Bengio said. "It's crazy to put those decisions into the hands of machines."

Even people who believe deeply in AI's potential, like Tesla CEO Elon Musk, have experienced the perils of giving automated systems too much authority. In 2018, Tesla was having trouble meeting production targets for its Model 3 sedan, in part because the company's factory machinery, which relied on a system of automated conveyor belts, kept malfunctioning.

After a number of frustrating production slowdowns, Musk stopped the belts, and brought in humans to replace the machines. Production sped up, and the company got back on track to meet its targets. Musk—who, bear in mind, believes that superintelligent AI will ultimately be a threat to human civilization—later admitted that he'd been wrong to place so much authority in the hands of machines.

"Excessive automation at Tesla was a mistake," he tweeted. "Humans are underrated."

To be clear, I am not advocating against using AI for important tasks. I am merely suggesting that we should be exceedingly cautious about giving machines more power than they can responsibly handle or putting algorithms in positions where its errors can harm innocent people.

Stronger government oversight could help here. In their book *Turning Point,* Brookings Institution researchers John R. Allen and Darrell M. West propose requiring companies and government agencies to submit "AI impact statements," similar to the environmental impact statements developers have to submit before starting a new project. These statements would outline the potential impacts of new automated systems on workers, and detail the steps being taken to mitigate those risks. In 2019, Senators Cory Booker and Ron Wyden, along with Representative Yvette Clarke, introduced something similar with the "Algorithmic Accountability Act," which would authorize the Federal Trade Commission to audit "highly sensitive automated decision systems," such as algorithms used for screening job candidates, for evidence of bias or flawed design.

Responsible tech companies can also help, by slowing down and considering how their new AI tools could be misused before making them publicly available. In 2019, OpenAI, the nonprofit AI lab, set a good example of responsible deployment when it withheld the full version of its new text generation algorithm, GPT-2. Experts had voiced concerns that GPT-2—which used AI to predict the next words in a sequence and could finish submitted samples of partial texts in an eerily humanlike way—could be used to spread fake news or computer-generated propaganda. So the organization released only a partial, less capable version until it could observe how it was being used in the wild. (It released the full version of GPT-2 nine months later, saying it had "no strong evidence of misuse so far.")

But we can't afford to wait for new laws, or depend on the moral compasses of AI makers, to start trying to prevent the reckless and irresponsible use of AI.

If you're a worker and your organization is implementing AI and automation without taking proper precautions, you can simply speak up. Make sure your manager understands the potential costs (financial, legal, and reputational) of an unforeseen error, and make the case for keeping humans involved in critical processes. Suggest a red-teaming exercise, in which you and your colleagues try to dream up and simulate all the ways an automated system could screw up or be misused. Or, as some manufacturing companies did in the 1970s and 1980s, form an "automation council," composed of workers in different divisions who can compare notes on how automated systems are working and present their findings to management.

If you live in a community where AI and automation are being used to violate people's privacy, punish disadvantaged people, or make high-stakes decisions about things like government benefits and housing, urge your local officials to show their work. If you can't be sure that the tools aren't having an adverse effect or if the data isn't available, research whether other communities have had issues with similar tools, and team up with civil liberties organizations if necessary. Interventions like these can be effective; in 2020, for example, the Chicago Police Department announced that it was dropping a contract to use facial-recognition technology developed by the controversial firm Clearview AI, after citizen activists and an ACLU lawsuit made the case that the technology could harm domestic violence survivors, undocumented immigrants, and other vulnerable groups.

If you're a leader deciding whether or not to use AI and automation in your organization, remember Mike Fowler and his algorithmic T-shirts. Then ensure that your algorithms are not going to introduce errors or perpetuate bias because of a flawed design or a biased training data set. (There are now "AI auditors" who can help evaluate algorithms for signs of these kinds of issues.) Tread carefully with third-party vendors and be skeptical of slick sales pitches. Involve workers in the process whenever possible.

And bear in mind, bosses, that human decision-makers, not bots, will bear the consequences of a flawed or premature AI deployment. If the chimp army destroys the office, metaphorically speaking, nobody's going to be mad at the chimps.

Rule 7

Build Big Nets and Small Webs

We are neither technologically advanced nor socially enlightened if we witness this disaster for tens of thousands without finding a solution. And by a solution, I mean a real and genuine alternative, providing the same living standards which were swept away by a force called progress, but which for some is destruction.
—Martin Luther King, Jr.,
in a speech about automation given to the
Transport Workers Union of America, 1961

On a frigid winter day a few years ago, I flew to Toronto, rented a car, and drove an hour north to a midsize Ontario city called Waterloo. I followed my GPS directions to a nondescript office park and pulled into a large lot with a small sign that read "Research in Motion—West Lot." A few cars were scattered throughout the lot, but when I arrived—a little after four P.M. on a weekday—it was mostly deserted.

A decade ago, this lot would have been full. At its peak, Research in Motion—maker of the BlackBerry—was one of the biggest names in technology, with $20 billion in annual sales and more than twenty thousand employees. RIM started here, and as it grew into a behemoth, its success turned Waterloo into a boomtown. That was before 2007, of course, when Apple introduced the iPhone and started RIM on a long, painful slide into obsolescence. As consumers flocked to iPhones and Android devices, RIM's BlackBerry sales cratered, its losses stacked up, and it was forced to lay off a huge chunk of its workforce.

The fall of RIM hit Waterloo hard, both economically and spiritually. The company had been the pride of the city, and had put it on the map as a global tech hub. Even as it became clear that the company's best days were behind it, people in town carried their BlackBerrys proudly, praying for a comeback.

It's not unusual for big companies to fail, nor is it unusual for the communities that rely on those companies to struggle for decades afterward. Just look at what has happened to Detroit since the peak of the U.S. auto industry in the 1960s, or how Rochester, New York, has fared since the collapse and bankruptcy of Kodak, its largest employer.

But unlike those industry towns, Waterloo didn't die. In fact, just the opposite. Most of the workers who got laid off from RIM quickly found other jobs. American tech companies like Google and Facebook swooped in to hire some of the company's workers, and local start-ups and big Canadian companies caught most of the rest. Today, the Waterloo economy is thriving, with a higher median household income and a lower unemployment rate than it had during RIM's heyday.

Part of Waterloo's quick recovery is related to the fact that many of the laid-off RIM workers were tech workers with in-demand skills. But tech skills don't explain the entire phenomenon, because RIM's nontechnical employees also got back on their feet quickly.

I spent a week in Waterloo, talking to local government officials, former BlackBerry workers, and community leaders about how the city recovered from BlackBerry's collapse. And I learned that there were two major factors behind the city's survival.

The first is what I came to call "big nets." Big nets are the large-scale programs and policies that soften the blow of sudden employment shocks. Canada's universal healthcare system functioned as a big net, as did its relatively generous unemployment benefits. In addition, Waterloo residents told me that the provincial government had stepped in at the first signs of trouble, doling out incentives for companies that were willing to hire the laid-off workers, and prioritizing keeping those workers from leaving the region.

The second factor in Waterloo's recovery was what I call "small webs"—the informal, local networks that support us during times of hardship. Waterloo is a tight-knit community full of small nets, and has a culture of generosity that goes back to the Mennonite roots of its original eighteenth-century settlers. When BlackBerry fell on hard times, these small nets activated. Communitech, a community technology education center and co-working space in town, gave free office space and other benefits to laid-off workers. Neighbors and friends traded job leads, and residents of the city organized job fairs and invited out-of-town employers to come.

One former BlackBerry employee, Dan Silivestru, told me that this collective response came naturally to people in town.

"It's a Mennonite barn-raising thing," Silivestru said. "When the problems at RIM started, everyone just put down their projects and went, 'Okay, time to help out.'" -

So far in this book, we've talked mostly about how we can prepare ourselves for the effects of technological change. But we should acknowledge, too, that no matter how hard we prepare, or how many human skills we develop, AI and automation might knock us off our feet anyway.

That's why I went to Waterloo. I wanted to figure out how the city survived a devastating technological change and kept itself from cascading into a permanent calamity. And I wanted to see if Waterloo's post-BlackBerry resilience held any lessons for other communities that might be disproportionately hit by the oncoming wave of AI and automation.

Let's be clear: not every community recovers as well as Waterloo. In northeast Ohio, where I grew up, the manufacturing economy that once supported hundreds of thousands of jobs in the region has been hollowed out by a combination of trade policy, automation, and political malpractice. Many of the workers who were laid off by GM, Ford, and other major employers in the region never landed on solid ground. They found lower-paying jobs, moved away from the region in search of better work, or simply dropped out of the workforce altogether. The loss of these jobs has been devastating for the area, and over the last two decades, the number of people living in poverty in my home county has doubled.

Similar tragedies have played out across the country, in communities where major, sustaining industries have disappeared. And many economists, technologists, and politicians who believe that AI and automation will displace many more workers have suggested big-net solutions in the form of sweeping policy changes and social programs that could soften the blow.

Historically, big nets have made it easier for societies to adapt to technological change. In Japan, for instance, a widespread labor practice called *shukko* helped soften the blow of major layoffs in the 1980s, as the country introduced robots into many of its factories. Under *shukko*, workers who were slated to be laid off could instead be temporarily "loaned" to other companies for as long as several years while the original employer found new work for them to do.

In Sweden, workers who lose their jobs to automation have benefited from the existence of groups known as "job-security councils." These councils—which have names I'd rather not try to pronounce, like Trygghetsrådet and Trygghetsstiftelsen—are private groups that cover workers in tens of thousands of companies. Employers pay into the councils, and when workers are laid off, the councils give them severance pay and personal job counselors, who help match them with open jobs and provide professional and emotional support as they look for other work.

Today, the most frequent big-net suggestion made by AI experts in the United States is universal basic income. Under a UBI plan, all adult citizens would receive a monthly, no-strings-attached cash grant, no matter their employment status or income. Several communities around the country are already testing small-scale UBI programs, and early results have been promising.

Some leaders, including Bill Gates and New York City mayor Bill DeBlasio, have proposed paying for expanded safety net programs by implementing a "robot tax," in which companies that deploy automated systems would pay an additional tax for each labor-displacing robot, comparable to a payroll tax for human workers. Others have proposed changing U.S. tax law, which currently incentivizes automation by taxing physical equipment, including computers and robots, at a lower rate than human labor, to give companies fewer reasons to rush to automate.

Many corporate chieftains have latched on to the idea of "reskilling" or "upskilling" programs, which would take workers with soon-to-be-obsolete skills, like driving trucks or operating forklifts, and train them to do other, more relevant work, like piloting drones or writing code. Companies like Amazon, AT&T, and JPMorgan Chase have rolled out ambitious retraining programs, and some state and local governments have set up their own workforce training and digital skills programs. But there is little evidence, so far, that these programs actually work at scale. Many companies find it easier to hire new employees than retrain existing ones, and some in-demand skills, such as data science, require specialized knowledge that can't be taught in a six-week seminar. A 2019 report by the World Economic Forum estimated that of the workers who will be fully displaced by automation in the next decade, only one in four can be successfully reskilled by private-sector programs.

Personally, I'm skeptical that the private sector will save us from a problem it is helping to cause. I'd favor a UBI-style plan coupled with Medicare for All and generous unemployment benefits for workers who are displaced by automation, similar to how the federal government stepped in with emergency cash transfers during the Covid-19 pandemic.

Whatever we do, it's inarguable that *any* collective action would be better than doing what we're currently doing to address automation-related economic pains on the federal level in the United States, which is essentially nothing.

In addition to big nets, we also need to think about the small webs we can create to support each other through this technological transition. Because in the absence of some fairly radical economic and policy changes, we're going to have to do a lot of this ourselves.

Our response to the Covid-19 crisis is a useful guide here. When the pandemic hit, state and local governments made up for the Trump administration's inept handling of the situation by gathering their own data, creating their own protocols, and building their own supply chains. Neighborhoods formed mutual aid networks to pool resources, arrange grocery deliveries and other assistance for needy and vulnerable residents, and help each other through financial distress. Donations flooded into food banks, worker relief funds, and small-business fundraising drives. People lent spare rooms to healthcare workers, and organized mask-sewing workshops.

Conscientious companies pitched in to create their own small webs, too. Airbnb, one of the hardest-hit Silicon Valley companies by the pandemic, was forced to lay off 25 percent of its staff after an unprecedented revenue drop. In addition to giving those employees generous severance pay, the company also helped them find new jobs by setting up an "alumni talent directory," with profiles and work samples of the laid-off employees, and turning its recruiting team (which wasn't doing much recruiting, after all) into a kind of makeshift outplacement firm. Executives from Accenture, Verizon, Lincoln Financial Group, and ServiceNow teamed up to build a platform to help laid-off workers connect with employers looking to fill open jobs, and signed up hundreds of participating companies.

Small webs don't have to be aimed purely at getting laid-off workers new jobs. Their benefits can be psychological—belonging to a religious organization or having a group meditation practice, for example, is a small web that can provide a sense of calm and purpose amid economic chaos. Volunteering at a local school, joining a book club, or merely cultivating new friendships—all of these are the kinds of small-web activities that can make us more resilient in the face of change.

Small webs can also help us learn how to use new technology for our benefit, and celebrate the instances in which new tools make our lives better.

One of my favorite small-web stories is that of the Rural Electrification Administration—the New Deal agency created by the Roosevelt administration in the 1930s to bring electric power to rural parts of the country—which held community-wide ceremonies each time it turned on the electricity for the first time in a new town. Getting electricity was a big, life-changing event for rural communities. It allowed farmers to dispense with heavy labor, extend the farming day by several hours, and produce bigger crop yields.

According to the historian David E. Nye, these ceremonies often turned into lively community parties, complete with speeches from local politicians and mock "funerals" in which an oil lamp was buried underground, to symbolize the death of an old technology and the arrival of a new one. At one such event in Kentucky in 1938, a preacher delivered a eulogy over the "casket" of a kerosene lamp, while the local Boy Scout troop played taps.

Today's new technologies rarely arrive to such a lively reception. But we could try to capture some of this celebratory communal feeling and come together as local communities to discuss and discover the implications of new technologies. You can imagine a block party, sponsored by a local chamber of commerce, to celebrate the arrival of 5G connectivity. Or an "AI fair" in which local families are invited to test out the latest medical robots, self-driving car prototypes, and machine learning programs being used by the companies in their area.

It's sometimes hard to remember, but technology once brought us together. It could do that again, if today's tech giants stop building technology that amplifies division and exacerbates inequality and start embracing their civic responsibility

But we don't have to wait for them. As a society, we can build more big nets to help people who are knocked off-balance by technological change. And as individuals, we can choose to create and strengthen small webs so that, if change comes to our doorstep, we'll have what we need to get by.

Rule 8

Learn Machine-Age Humanities

*We're training people to do machine things. We shouldn't
be doing that. We should be training people in uniquely
human capabilities.*

—Paul Daugherty,
chief technology and innovation officer, Accenture

Since I started writing about AI and automation, a lot of nervous parents have asked me what subjects their kids should
study in order to be prepared for the future.

For a long time, I didn't have a good answer. Because while
I'm confident that the most valuable skills for the future are
lowercase-*h* humanities—those surprising, social, and scarce
abilities that we've been talking about—I'm much less confident that just studying the traditional, capital-*H* Humanities
subjects in school will get them there. Is the average anthro-
pology major likely to be more socially adept than the average
engineering major? Does reading *Beowulf* make you better at

handling surprises, or developing scarce talents, than learning Bayesian statistics?

Many ideas have been proposed and tested for bringing our educational system into the twenty-first century, including personalized curricula, massive open online courses (MOOCs), and "lifelong learning" adult education programs. But few of them have been adequately tested, and all of the ideas deal primarily with *how* we should teach people, leaving open the question of *what* we should teach them. And because all of them are geared toward reforming our current model of education—emphasizing certain subjects, de-emphasizing others, adjusting class sizes, and updating pedagogical methods—there's a lot they're missing.

Recently, I decided to make my own list of essential skills for the future. I call them "machine-age humanities" because, while they're not strictly technical skills, they're not exactly classic humanities disciplines like philosophy or Russian literature, either.

Instead, they're practical skills that I think can help everyone—from young kids to adults—maximize their advantages over machines.

Attention Guarding

Daniel Goleman, the psychologist who popularized the term "emotional intelligence," believes that focus—the ability to direct one's own attention—will be a key skill of the future. He writes that being able to focus, and tune out external distractions, will be helpful for navigating a fast-changing future, and dealing with the ups and downs we're likely to experience as a result of technological change.

"Those who focus best are relatively immune to emotional turbulence, more able to stay unflappable in a crisis and to keep on an even keel despite life's emotional waves," Goleman writes.

I prefer the term "attention guarding" to "focusing," since it acknowledges that today, when most of us fight distraction, what we're actually doing is defending our attention from an assault by various external forces—social media apps, breaking news alerts, a cavalcade of texts and emails—that are attempting to distract and divert us.

There are established ways to train our brains to better guard our attention. Meditation is one; studies have shown that periods of meditation even as short as eight minutes can reduce mind-wandering. Breathing exercises, nature walks, and prayer can also help. For me, the best attention-guarding ritual of all is reading—sitting down to read physical, printed books for long stretches of time, with my phone sequestered somewhere far away. But we could use more research into attention-guarding tactics, especially given the enormous amount of brainpower and money being spent to distract us.

Guarding attention is typically thought of as a productivity hack—a way to get more done, with less distraction. But there are noneconomic reasons to practice keeping our attention away from the forces trying to capture and redirect it. Sustained focus is how we develop new skills and connect with other people. It's how we learn about ourselves and construct a positive identity that can withstand influence from machines. After all, as the historian and author Yuval Noah Harari writes, "If the algorithms understand what's happening within you better than you understand it, authority will shift to them."

Room Reading

Recently, I went to a talk by Jed Kolko, the chief economist of Indeed.com, who made an unexpected prediction about a group of people who may be well-prepared for the future. LGBTQ people who had spent time in the closet, he said, might fare especially well in the age of AI and automation, because many of them are experienced in the kind of delicate social maneuvering that requires a high level of emotional intelligence.

"The kind of skill that one gets from being in the closet— the ability to read a room—that's not a skill that shows up anywhere in a skills inventory, but ends up being the kind of skill that can be valuable in all kinds of workplaces," Kolko said.

I'd take Kolko's prediction a step further and argue that women and racial minorities—many of whom are forced to code-switch and modulate their behavior in workplaces dominated by white men every day—will also be well positioned for the future. The same instinct that leads a woman executive to soften her tone in order to avoid being seen as aggressive, or that informs a Black employee that she should modulate away from African American Vernacular English while presenting to a group, may also give them a leg up in fields that require high levels of social perception.

Of course, it would be much better to live in a more equitable society, where women and minorities weren't required to manage their self-presentation so carefully. But the machine age may present a silver lining for people who have gotten good at quickly assessing the biases and prejudices of others.

And those of us who don't bear the burden of code switching and room reading should try to cultivate these skills in other ways, because we'll need them.

Resting

One of my favorite social media follows is an Instagram account called "The Nap Ministry."

It's run by Tricia Hersey, a Black performance artist and poet from Atlanta, Georgia. Several years ago, while studying in divinity school during the early days of what became the Black Lives Matter movement, Hersey found herself exhausted and worn down by both her studies and the well-publicized videos of police brutality against Black people. She decided to start taking daily naps. And after observing the effect these naps were having on her mental health, she dubbed herself the "Nap Bishop," and started The Nap Ministry, with the goal of teaching other people—and especially other emotionally exhausted Black people—about the transformative potential of taking naps.

"Rest is productive," Hersey told one interviewer. "When you are resting, you are being productive. I'm trying to reframe rest and deprogram people around the concept that if you aren't 'doing something' in the classic sense, then you're not worthy."

Hersey believes that taking naps and relaxing is about more than just self-care—it is an act of resistance against the pressures of white supremacy and capitalism, and a move to reclaim Black bodies from hustle culture. Her Instagram account is full of inspirational quotes like "Rest is a liberation practice" and "You are not a machine. Stop grinding."

Even though I am not the target audience for Hersey's activism, I deeply appreciate her reframing of resting our bodies as a social justice issue, and a necessary skill for those who need the energy to resist oppression and fight for a fairer future.

We generally stop incorporating naptime into education after early childhood. But resting—turning off our brains, recharging our bodies—is an increasingly useful skill for people of all ages. It helps prevent burnout and exhaustion, allows us to step back and look at the bigger picture, and helps us step off the hamster wheel of productivity and reconnect with the most human parts of ourselves. And many of us, including me, could use a refresher course.

In the old economy, when our value was mostly predicated on our physical labor, midday rest was often seen as an indulgent luxury. But in the new economy, when more creative and human skills are what will differentiate us from machines, we should reframe our attitude toward rest, viewing it as a critical survival skill. The science is quite clear on the link between rest and all kinds of human function. Studies conducted by neuroscientists at the Walter Reed Army Institute of Research and other top institutions have found that chronic sleep deprivation impairs our moral judgment, lowers our emotional intelligence, and harms our interpersonal communication skills. (To say nothing of its risks to our physical health.)

In addition to developing our own napping skills, we should also push for structural changes that could alleviate burnout and overwork on a larger scale. This is already happening in other countries. In Japan, a 2019 law limited worker overtime to forty-five hours per month, with financial penalties for companies that ignored the limits. A French law that went into effect in 2017 gives workers the "right to disconnect," and legally protects them from being required to respond to emails after six P.M. In America, some companies have begun enforcing mandatory vacation policies, and shutting off company-wide emails on the weekend.

Some schools are also experimenting with educating students about the value of resting themselves. At Harvard, incoming freshmen are now required to take an online course known as "Sleep 101," adapted from a popular seminar taught by the leading sleep researcher Charles Czeisler, before ever setting foot on campus. Brown, Stanford, and NYU also offer their own, optional courses in sleep studies.

But these kinds of courses can't be reserved for elite college students. In the automated future, as more of our contributions come from big breakthroughs, inspired ideas, and emotional aptitude, being well-rested is going to become even more critical.

Digital Discernment

As a tech columnist who writes about social media, I've spent a lot of time over the past few years reporting on misinformation and conspiracy theories. And I've noticed, as I'm sure you have, that even very intelligent people have been struggling to figure out what's true and false these days.

This is not an accident. Billions of people get their news and information from social networks like Facebook, Twitter, and YouTube, all of which use algorithms that reward information for being engaging, regardless of whether or not it's true. These platforms design ads to look as similar to organic posts as possible—meaning that most users quickly scrolling through their feeds can't tell the difference between a paid message and an independent one. And in the rare instances when these platforms do fact-check a post—putting a link to a World Health Organization page about vaccine safety next to a post from a rabid anti-vaxxer, say—they've conditioned users to mistrust mainstream authority so thoroughly that the fact-check itself often becomes fodder for more conspiracy theories.

I don't love the popular phrase "media literacy," which implies that people can be taught a single, correct way to synthesize and interpret news and information sources, many of which conflict and collide with each other—and some of which are deliberately designed by bad-faith media hackers to fool their audiences and manipulate public opinion.

Instead, I prefer to talk about "digital discernment," which reflects the fact that learning to navigate our way through a hazy, muddled online information ecosystem is a continuous, lifelong process that changes as technology shifts, and as media manipulators adapt to new tools and platforms.

Our lack of digital discernment is becoming a real societal problem. In 2015, a group of Stanford researchers conducted a study of "civic online reasoning," in which they gave basic news literacy tests to more than seven thousand middle school, high school, and college students. In one test, participants were shown an article on financial planning that was sponsored by a bank and written by a financial executive, and asked whether it was likely a credible, objective source. Another test asked participants to evaluate two similar-looking Facebook posts—one from the official Fox News account, and another from an impostor page—and indicate which one was authentic. The results were shockingly bad. More than 80 percent of participants mistook a native ad—a story paid for by an advertiser, identified by the label "sponsored content"—for a real news story. More than 30 percent found a fake Fox News Twitter account more credible than the real one.

"In every case and at every level, we were taken aback by students' lack of preparation," the researchers wrote.

Digital discernment isn't just a problem for young people. In fact, one study found that during the 2016 election, people sixty-five and older were seven times more likely to share internet-based misinformation than younger people. And while debunking internet misinformation is already hard, it's going to get even harder in the coming years, with the rise of algorithmically generated text, realistic conversational AI, and synthetic video ("deepfakes") produced with the help of machine learning.

There is no perfect digital discernment solution, but researchers have made some headway. In a 2018 report for the nonprofit organization Data & Society, Monica Bulger and Patrick Davison write that while media literacy programs have some limitations, certain types of interventions can be effective. They mention the example of #CharlottesvilleCurriculum, a hashtag that trended on Twitter after the deadly Unite the Right rally of white nationalists in Charlottesville, Virginia, in 2017. Following the rally, which resulted in a rash of hyperpartisan misinformation, educators and organizations such as the Anti-Defamation League used the hashtag to share advice for fostering constructive classroom dialogue about race, bias, and tolerance.

This is a good start. But we desperately need more research into what kinds of interventions actually work—not just stopping people from being duped by misinformation in the first place, but pulling them back to reality once they've started believing in a conspiracy theory or falling for hoaxes. In this upside-down, epistemically confused environment, being able to tell fact from fiction will be a human superpower. Digital discernment will allow people to filter information more effectively, to avoid getting duped by hoaxers and charlatans, and see through the fog of the modern information wars.

Analog Ethics

Frank Chen, a venture capitalist who invests in AI start-ups, recommends an unconventional book to people who ask him what skills will be valuable in the future. The book, *All I Really Need to Know I Learned in Kindergarten*, was written in 1986 by a minister, Robert Fulghum, and it's full of simple-sounding life advice, like "share everything," "play fair," and "clean up after your own mess."

Chen believes that these skills the elementary, pre-literate skills of treating other people well, acting ethically, and behaving in prosocial ways, all of which I consider "analog ethics"—are badly needed for an age in which our value will come from our ability to relate to other people. He writes:

> While I know that we'll need to layer on top of that foundation a set of practical and technical know-how, I agree with [Fulghum] that a foundation rich in EQ [emotional quotient] and compassion and imagination and creativity is the perfect springboard to prepare

people—the doctors with the best bedside manner, the sales reps solving my actual problems, crisis counselors who really understand when we're in crisis—for a machine-learning powered future in which humans and algorithms are better together.

Research has indicated that teaching analog ethics can be effective. One 2015 study that tracked children from kindergarten through young adulthood found that people who had developed strong prosocial, noncognitive skills—traits like positivity, empathy, and regulating one's own emotions—were more likely to be successful as adults. Another study in 2017 found that kids who participated in "social-emotional" learning programs were more likely to graduate from college, were arrested less frequently as adults, and had fewer diagnoses of mental health disorders, even when variables like race, socioeconomic status, and school location were controlled for.

For younger kids, of course, basic skills like sharing, playing fair, and apologizing never left the curriculum. But schools are now starting to explicitly design programs around cultivating kindness. The Kindness Curriculum, a set of instructional materials developed by the Center for Healthy Minds at the University of Wisconsin–Madison, helps preschoolers learn basic mindfulness skills that can help them understand their own emotions and the emotions of others. And Roots of Empathy, a program developed by the Canadian educator Mary Gordon to help students develop empathy and emotional literacy, is being used by schools in fourteen countries, including the United States, South Korea, and Germany.

Older students, too, are experimenting with revisiting analog ethics. At Stanford, for example, students can take a seminar called "Becoming Kinder," which teaches them about the psychology of altruistic behavior. At NYU, an undergraduate course called "The Real World" is teaching students a critical skill of the future—the ability to cope with change—by conducting simulated problem-solving drills. At Duke, Pittsburgh, and other top medical schools, oncology fellows can sign up for "Oncotalk," a specialized communications course that teaches them how to have difficult conversations with their cancer patients.

These efforts are all a good start, and more analog ethics teaching is deeply necessary—not just to improve people's personal lives, but to equip them for a future in which our social and emotional skills will be some of our most precious assets.

Consequentialism

Some of the most valuable skills in the future will involve thinking about the downstream consequences of AI and machine learning and understanding the effects these systems are likely to have when they're unleashed into society.

We now know some of the unintended consequences of planetary-scale AI systems like Facebook and YouTube, and how the engineers and executives who conceived those systems failed to appreciate the ways the products they built could be misused, exploited, and gamed. Most of these systems, I believe, were not intentionally designed to create harm. Instead, I think their founders and engineers were idealists who thought that having good intentions mattered more than producing good outcomes.

Today, in part because of these blind spots—and the billions of dollars their companies have had to spend fixing their mistakes—the demand for people who can spot the flaws in technological systems *before* they cause catastrophic problems has grown. Big tech companies are hiring people with backgrounds in fields like law enforcement, cybersecurity, and public policy, who have both the real-world experience and the consequentialist imaginations to analyze new products and imagine all the possible harms they could enable.

There will be a much bigger need for these people in the future, and not all of them will be engineers. Some of them may just be people who understand human psychology, or risk and probability. (Jack Dorsey, the chief executive of Twitter, has said that he regrets not hiring a game theorist and a behavioral economist during Twitter's early days, to help the company understand the ways bad actors might abuse their systems.)

Consequentialist thinking will be useful outside of tech, too, as AI moves into more industries and creates more opportunities for error. Doctors and nurses will need to understand the strengths and weaknesses in the tools used for diagnostic imaging and anticipate how they could produce faulty readings. Lawyers will need to be able to peer inside the algorithms used by courts and law enforcement agencies and see how they could result in biased decisions. Human rights activists will need to know how things like facial-recognition AI could be used to surveil and target vulnerable populations.

One way to instill consequentialist thinking would be by formalizing it as part of a standard STEM curriculum or turning it into a professional rite of passage. In Canada, when you graduate from engineering school, you're invited to take part in a ceremony called the Ritual of the Calling of an Engineer, which dates back to the 1920s. During the ceremony, graduates are each presented with an iron ring, worn on the pinkie finger, that is supposed to remind them of their responsibilities to serve the public good. They then recite an oath, which begins with a pledge that they will "not henceforward suffer or pass, or be privy to the passing of, Bad Workmanship or Faulty Material."

Imagine if software engineers at Facebook and YouTube were required to undergo a similar ceremony before shipping their first feature or training their first neural network. Would it solve all of society's problems? Of course not. But could it remind them of the stakes of their work, and the need to be mindful of the vulnerabilities of their users? It's certainly possible.

Rule 9

Arm the Rebels

We are all afraid—for our confidence, for the future, for the world. That is the nature of the human imagination. Yet every man, every civilization, has gone forward because of its engagement with what it has set itself to do.

—Jacob Bronowski

Nearly two centuries ago, a world-weary twenty-seven-year-old decided to take a break from technology.

The man was from Concord, Massachusetts, the epicenter of the American Industrial Revolution. His family owned a successful pencil factory, which afforded him a comfortable existence. But factory life didn't suit him. After college, he'd gotten interested in Transcendentalism, a new movement of New England writers and philosophers who had become disillusioned with modernity, which they believed was dehumanizing people and turning them into bland conformists.

Eventually, he decided to leave the industrialized world behind. He built a small, bare-bones house on a patch of land next to a pond, got rid of his possessions, and went to live there.

The man, Henry David Thoreau, immortalized his trip to the pond in *Walden*—a book that changed the way generations of Americans have looked at the trade-offs of technological progress. *Walden* became famous for Thoreau's depictions of nature and ruminations on the simple life, but the book was also a scathing anti-technological rant. Thoreau plainly hated technology, and resented the hype surrounding new gadgets like the telegraph, which he saw as nothing but a distraction from man's true purpose.

"We are in great haste to construct a magnetic telegraph from Maine to Texas; but Maine and Texas, it may be, have nothing important to communicate," he wrote in 1854. "As if the main object were to talk fast and not to talk sensibly."

Most people know Thoreau's story. But many fewer people know that on July 4, 1845—coincidentally, the same day Thoreau was moving to Walden Pond—a labor activist named Sarah Bagley was giving a speech that would change the tra jectory of technological progress much more directly than anything Thoreau ever wrote.

Bagley lived in Lowell, Massachusetts, and she'd grown up as a "Lowell girl"—one of the many young working-class women who worked in the local textile factories. Like Thoreau, Bagley had become disenchanted with industrial culture, but for much different reasons. She was a worker, not the child of a well-off industrialist, and she'd seen firsthand how bad life in the factories was. She'd experienced pay cuts, long hours, and inhumane working conditions, and it infuriated her that industrial barons were getting wealthy at the expense of workers.

Instead of retreating into nature, Bagley became a labor organizer. She began submitting pro-worker articles to a local magazine, and eventually organized a workers' rights organization called the Lowell Female Labor Reform Association. Local labor leaders noticed her efforts and invited her to give a speech at an Independence Day gathering of workers in Woburn, Massachusetts.

The speech was a big deal. Roughly two thousand people gathered in an outdoor grove to hear Bagley speak about the injustices of the Industrial Age. She lampooned the factory owners, calling them the "mushroom aristocracy of New England." She vowed to join men's labor unions in their fight for a ten-hour workday and other worker protections. And she defended the Lowell girls, saying, "Our rights cannot be trampled upon with impunity."

Bagley's speech killed, and energized a labor movement whose spirits had been deflated by harsh opposition. One local newspaper called her a "lady of superior talents and accomplishments." When she finished, the paper reported, she was greeted with the "loud and unanimous huzzas of the deep moved throng."

I'm telling you these stories not because I thought this book needed to end with another dose of nineteenth-century history, but because they illustrate one of the most important choices we face as we prepare for our technological future.

In many ways, the world today looks a lot like it did in 1845. New, powerful machines have revolutionized industries, destabilized legacy institutions, and changed the fabric of civic life. Workers are worried about becoming obsolete, and parents are worried about what new technologies are doing to their children. Unregulated capitalism has created an extraordinary amount of new wealth, but workers' lives aren't necessarily getting better. Society is fractured along lines of race, class, and geography, and politicians are warning about the dangers of rising inequality and corporate corruption.

In the face of these challenges, we have two options.

We can do what Thoreau did—throw our hands up, unplug our devices, opt out of modernity and retreat into the wilderness. Or we can do what Sarah Bagley did. We can step into the conversation, learn the details of the power structures that are shaping technological adoption, and bend those structures toward a better, fairer future.

Personally, I'm on Team Bagley. I think we have a moral duty to fight *for* people, rather than simply fighting against machines, and I believe that for those of us who aren't tech workers, that duty extends to supporting ethical technologists who are working to make AI and automation a liberating force rather than just a vehicle for wealth creation.

I call this strategy "arming the rebels," not because I think resisting technological exploitation should involve violence of any kind, but because I think it's important to support the people fighting for ethics and transparency inside our most powerful tech institutions by giving them ammunition in the form of tools, data, and emotional support.

On a practical level, I think this strategy is likely to be more effective than trying to tear down these institutions altogether. History shows us that those who simply oppose technology, without offering a vision of how it could be made better and more equitable, generally lose. The Luddites earned their place in the history books by breaking their weaving machines, but they didn't reverse the effects of industrialization. Skeptics who pooh-poohed the idea of space travel in the mid-twentieth century were complaining into a void, but those who actually engaged with the project—including underappreciated heroes like Katherine Johnson, Dorothy Vaughan, and Mary Jackson, the Black female NASA engineers whose contributions to the space race were memorialized in the book and film *Hidden Figures*—got to shape one of our nation's greatest technical accomplishments. The people who lamented the early days of the internet might have been satisfied on their moral high ground, but they also missed the chance to help shape the online spaces that would affect billions of people's lives in the coming decades.

I routinely get emails and DMs from today's Sarah Bagleys—rank-and-file workers at Facebook, Amazon, Google, and other tech giants who tell me that they're horrified by some of the tools their companies are building, their workplace practices, and their failures to contain the harms resulting from the use of their products. These people believe that they can be most effective as ethical advocates from the inside, but they're grateful for the efforts of journalists, researchers, and activists who are pushing from the outside, adding more voices to the chorus calling for change.

And outside the big tech companies, there are plenty of righteous machine-makers we can support and learn from.

These are people like Jazmyn Latimer, a product designer who works with the nonprofit group Code for America. Several years ago, Latimer got an idea for an app called "Clear My

Record," which would use automated software to help eligible people convicted of past criminal offenses to apply to have their convictions expunged from their records. The app has been used in California to clear more than eight thousand low-level drug offenses, giving thousands of formerly incarcerated people a clean slate.

Or Rohan Pavuluri, a twenty-three-year-old Harvard graduate who started a legal aid nonprofit called Upsolve in 2016. The organization uses automated software to help low-income Americans file for Chapter 7 bankruptcy, a process that allows them to shed burdensome debt obligations and get a fresh financial start. So far, the service has helped families clear more than $120 million in debt.

Or Joy Buolamwini and Timnit Gebru, two AI researchers who studied three leading facial-recognition algorithms, and found that all three were substantially less accurate when trying to classify darker-skinned faces than lighter-skinned faces. The study led several major tech firms to reexamine their AI for evidence of bias, and pledge to use more racially diverse data sets to train their machine learning models.

Or Sasha Costanza-Chock, a nonbinary, transgender media scholar and MIT professor who has promoted the concept of "design justice"—an approach to product design that explicitly tries to dismantle structural injustice and center the needs of marginalized people. Costanza-Chock has been a leader in the movement to ban facial-recognition technology and oppose the use of tools that harm vulnerable people, such as the millimeter wave detection machines used at airports, which force TSA agents to pick a binary male/female gender before scanning a passenger's body.

I've featured people like these for the last several years in my annual "Good Tech Awards" column, because I think we need to create incentives—even small ones, like being mentioned in a newspaper story—for people to build technology that helps people at scale, rather than just making money for themselves and their investors.

As we fight to shape today's technological landscape, I think we have a special obligation to fight for the people who stand to lose the most from AI and automation, including historically marginalized communities and people who don't have much of a safety net.

I also think we need to resist the urge to push the AI conversation too far into the future. I've always loved the concept of the "adjacent possible," a term coined by the evolutionary biologist Stuart Kauffman to describe the way biological organisms evolve in gradual, incremental steps.

The adjacent possible is a useful concept to apply to the world of technology, because it takes us out of the realm of sci-fi and narrows our scope to more realistic outcomes. A world in which robots flawlessly perform all human labor, freeing us all up to make art and play video games every day, is probably not part of the adjacent possible. But a world in which we use machine intelligence to reduce carbon emissions, find cures for rare diseases, and improve government services for low-income families might be.

It's on us—the people who love technology but worry about its use—to explore this adjacent possible and push for the best version of it.

It's also important not to get too discouraged, and to remember, despite all of our worries, that AI and automation could be *unbelievably good* for humankind, if we do it right. A world filled with AI could also be filled with human creativity, meaningful work, and strong communities. And it's worth reminding ourselves that, historically, technological shocks have been followed by social progress, even if it's taken a while. The worker unrest of the Industrial Revolution led to labor reforms and the first institutionalized protections for workers. Worries about automation in the middle of the twentieth century strengthened the middle class by expanding the power of labor unions. The rise of the "gig economy" in the first decade of the twenty-first century has already created a groundswell of organizing energy to protect contract workers from exploitation.

Look, I don't judge people for wanting to unplug their devices and flee to the hills. And I'm certainly not opposed to adopting a balanced lifestyle that puts technology in its proper place. But technological abstinence is not the answer to our problems, and I believe that we have to engage with potentially harmful systems in order to influence their trajectory.

It's easy to see how AI could tear us apart. But it's also easy to see how it could unite us. Technology can force us to study ourselves, and figure out our own strengths and limitations. Machines can foster resilience and creativity, as we come up with new and creative ways to stay ahead. And AI and automation could bring us together, armed with new superpowers, to solve some of our biggest problems.

But none of this will happen without us. The future is not a spectator sport, and AI is too important to be left to the billionaires and bot builders. We have to join the fight, too.

•   •   •

On February 21, 1846, less than a year after she thrilled a crowd with her call for workers' rights, Sarah Bagley made history again.

An associate of Samuel Morse, the famed inventor of the electric telegraph, had come through Lowell to inquire about a new planned telegraph line from Boston to New York. The line would need a depot in Lowell, and Morse was looking for a qualified person to run it. He asked Bagley if she was interested in the job.

Bagley had no experience operating a telegraph. She'd been a mill worker and a labor organizer, and the telegraph was a new, state-of-the-art gadget that required specialized training. Few women had operated telegraphs before, and some men doubted it was possible. ("Can a woman keep a secret?" one local newspaper wondered.) In addition, the telegraph's future was far from certain.

But Bagley was a risk taker, and she liked a challenge. So, she said yes. She accepted a salary of $400 a year, spent several weeks studying the mechanics of the telegraph, and went to work.

Bagley had no real need for a career change. She was already a legend in the New England labor movement, and she could have coasted on her reputation for years.

But she didn't want to spend the rest of her life talking about the previous chapter of history. She wanted to write the next one.

Acknowledgments

Book-writing cannot yet be automated; as such, there are a number of humans to thank for their kind assistance in this project.

I am grateful to my colleagues at the *Times* for their support, advice, and forbearance: A. G. Sulzberger, Dean Baquet, Joe Kahn, Rebecca Blumenstein, Sam Dolnick, Ellen Pollock, Pui-Wing Tam, Joe Plambeck, Mike Isaac, Nellie Bowles, Natalie Kitroeff, Cade Metz, Kara Swisher, Lisa Tobin, Michael Barbaro, Andy Mills, Larissa Anderson, Wendy Dorr, Julia Longoria, Sindhu Gnanasambandan, and countless more people who helped in ways large and small.

I am grateful to my editor, Ben Greenberg, for seeing the potential in this book and steering it to completion, as well as to the rest of the team at Random House, including Ayelet Gruenspecht, Molly Turpin, and Greg Kubie. Sloan Harris and Kari Stuart of ICM Partners were patient and thoughtful counselors, as always. Rachel Gogel designed another knockout cover.

I am grateful to all the sources and experts who sat for interviews, sent reading suggestions, and reviewed early drafts, with special thanks to Roy Bahat, Catherine Price, Jessica Alter, and A. J. Jacobs.

I am grateful to Tori Jueds and the students of Westtown School, who gave valuable early-stage feedback.

I am grateful to friends and family members who provided edits, office space, moral support, and various forms of assistance, including Paul Roose, Anne Lawrence, Nikole Yinger, Aaron Freedman, Alexis Madrigal, Andrew Marantz, Sarah Lustbader, Ariel Werner, Ari Savitzky, Kate Lee, Caroline Landau, Alex Goldberg, and many others I am surely forgetting.

I am grateful to my immediate family—Diana Roose, Carl Roose, and Julia Slocum—for their love and support, even as I broke my "no more books" pledge. And I am grateful for the memory of two late relatives who encouraged me to become a writer in the first place: my father, Kirk Roose, who died in 2018, and my grandmother, Gretchen Roose, who died in 2020, while this book was being edited, and who would surely have given out copies to everyone she knew.

Finally, I am grateful beyond measure to my partner and spouse, Tovah Ackerman, who makes me excited to be a human every day.

Appendix

Making a Futureproof Plan

Most of the advice in this book has taken the form of broad, generally applicable principles that could be useful to lots of different types of people, no matter their situation. That was deliberate—and if general principles are what you're after, and you feel like you've gotten what you needed to get out of this book, great!

If, however, you're still craving a little more specificity, then I'd recommend spending some time making a plan.

What this plan looks like is up to you. Some people like lists of short-term, bite-sized goals. Others like focusing on longer-term transformations. Some people don't like goal setting at all, and would rather have one or two things they simply remind themselves of every day. (I have a friend who keeps motivational Post-it notes on his computer monitor that say things like "drink more water" and "don't be a jerk on Twitter.")

Personally, I like achievable, short-term goals. So for Rule 1 ("Be Surprising, Social, and Scarce") I created a three-by-three matrix, and filled each box with a different goal for each of three domains of my life (home, work, community). Mine looks like this:

	SURPRISING	SOCIAL	SCARCE
HOME	Bring home flowers for no reason.	Call an old friend I haven't talked to in years.	Read a book nobody else I know is reading.
WORK	Write a story for a section other than tech.	Organize a virtual happy hour.	Master NewsWhip (a social media analytics tool that few other *Times* reporters use regularly).
COMMUNITY	Go out for a night with the "pothole vigilantes" (a group of Oakland residents who fill potholes under cover of darkness without notifying the city).	Host a dinner party for neighbors.	Sign up for an emergency preparedness class.

I also created some goals based on the other eight rules in the book. Here's what my current list looks like:

Rule 2: Resist Machine Drift

• Turn off YouTube recommendations, and shop offline whenever possible.

- Protect "human hour."
- Meditate daily.

Rule 3: Demote Your Devices

- Keep phone screen time below 1.5 hours per day.
- No email on Sundays.
- Repeat Catherine Price's thirty-day program once a year as necessary.

Rule 4: Leave Handprints

- Send one handwritten note per week.
- Mentor a journalism student.
- Send detailed positive feedback to colleagues whose work I appreciate.

Rule 5: Don't Be an Endpoint

- Stop checking Stela. (Stela is the *Times*'s internal analytics dashboard; it's a useful tool for editors, but it can also make me too aware of how many views my stories are getting, which can have the effect of making me want to write only stories that I know will get lots of views.)
- Reserve Friday afternoons for reading and developing new sources.
- Go to the office at least three days a week (pandemic permitting).

Rule 6: Treat AI Like a Chimp Army

- Investigate how algorithms are affecting the criminal justice system in the Bay Area.

- Take an online course in machine learning.
- Meet with the *Times*'s chief data scientist to learn how we use algorithmic recommendations in our app and on our website.

Rule 7: Build Big Nets and Small Webs

- Organize a block party for neighbors.
- Start attending a Quaker meeting.
- Get more involved in the News Guild, the *Times*'s labor union.

Rule 8: Learn Machine-Age Humanities

- Pay more compliments.
- Read stories in full before sharing them on social media.
- Take at least one midday nap a week.

Rule 9: Arm the Rebels

- Meet with Civic Signals, a group of scholars and activists who are working to make digital platforms function more like public spaces.
- Increase the number of nonwhite, non-male sources I quote in my stories. (I stole this idea from my colleague Ben Casselman, who does an annual "diversity audit" of his own articles to make sure women and people of color are adequately represented among his named sources.)
- Donate to Fast Forward, an accelerator for tech nonprofits working to solve major societal problems.

These goals are obviously not one-size-fits-all, and yours will probably look quite different. You'll also notice that many of my goals aren't related to AI or automation at all—they're more like general self-help tips. That's kind of the point. If the way to survive technological change is to become more human, then much of what we need to do is going to be repairing and restoring the basic skills we may have let decay.

For me, a futureproof plan is a way of holding myself accountable and reminding myself that my daily choices add up. It's a way of measuring my own progress toward becoming more human. All I ask is that if you do create your own plan, make sure it addresses your entire life, and not just your work. Futureproofing is about reclaiming control of our minds and our human agency, not just keeping our jobs.

Reading List

The bad thing about writing a book about AI and automation is that a lot of great writers got there first. That is also the great thing. Many authors I admire have already taken on the challenge of explaining some piece of the technological future, and readers hoping to learn more about the topic have a wealth of great books to choose from.

Here are a few of the books that most informed my thinking on the subjects of AI, automation, and the contours of the future. If you're looking to start your own robot-themed bookshelf, I'd recommend starting here.

Artificial Unintelligence by Meredith Broussard (2018). Broussard, an experienced data journalist and NYU professor, is a savvy guide to the foibles and limitations of AI, and her book is a forceful argument against what she calls "technochauvinism."

The Second Machine Age by Erik Brynjolfsson and Andrew McAfee (2014). This book by two MIT professors was years ahead of its time. I find myself going back to it frequently.

Humans Are Underrated by Geoff Colvin (2015). Colvin, a longtime writer and editor at *Fortune,* makes a compelling case for the economic value of human skills.

Human + Machine by Paul R. Daugherty and H. James Wilson (2018). A largely optimistic insider's view on corporate automation from two in-house AI and automation experts at Accenture, the consulting firm.

Mind over Machine by Hubert L. Dreyfus and Stuart E. Dreyfus (1985). Written by a father-son pair of professors in philosophy (Hubert) and engineering (Stuart), this book was an early attempt to answer questions about the limits of digital technology.

Rise of the Robots by Martin Ford (2015). A lively, accessible, and slightly terrifying view of the robot age, from a journalist who has been following it longer than most.

The Technology Trap by Carl Benedikt Frey (2019). An Oxford economist's robust history of technological change, with new research and assumption-challenging conclusions.

AI Superpowers by Kai-Fu Lee (2018). Lee, an experienced AI leader and venture capitalist, gave me new language to use in talking about AI and humanity, and provided a valuable window onto the Chinese AI market.

Machines of Loving Grace by John Markoff (2015). Markoff, a legendary tech journalist and my former colleague at the

Times, is an essential guide to the world of AI, and an expositor of the people and philosophy behind its design.

Forces of Production by David F. Noble (1984). This book, an examination of the automation landscape after World War II, was essential to my thinking on the culture of industrial automation. Noble is a good writer and an excellent historian, and makes a persuasive case that automation has been used not just as a tool to make companies more productive, but to exert authority over workers.

Technopoly by Neil Postman (1992). A classic work by one of the great technological critics, who anticipated many of the ways that technology could challenge our humanity.

How to Break Up with Your Phone by Catherine Price (2018). This book, which was written by my phone detox coach, is a life-changer. Catherine's thirty-day phone detox plan revolutionized my relationship with my phone, and got me thinking about the control I'd surrendered to machines. I've bought more copies of this book for friends and family members than I can count.

Inside the Robot Kingdom by Frederik L. Schodt (1988). Another fascinating 1980s automation book, this time about Japan and its culture of relentless factory roboticization.

You Look Like a Thing and I Love You by Janelle Shane (2019). A great primer, written by a prominent AI researcher, and one of the only books about machine learning that has ever made me laugh.

Future Shock by Alvin Toffler (1970). The book that kicked off the futurist craze, and still one of the best examples of writing about the psychological effects of technological change.

The Human Use of Human Beings by Norbert Wiener (1950). An examination of the morality of machines, written by one of my all-time favorite technological thinkers.

In the Age of the Smart Machine by Shoshana Zuboff (1988). Zuboff is better known these days as the author of *Surveillance Capitalism*, but her earlier book was a prescient look at the future of work during the first IT boom of the 1980s.

Notes

Introduction

xviii **I got my first glimpse** Kevin Roose, "The Hidden Automation Agenda of the Davos Elite," *New York Times*, January 25, 2019.

xx **Aristotle mused that automated weavers** Sean Carroll, "Aristotle on Household Robots," *Discover*, September 28, 2010.

xx **In 1928, *The New York Times* ran an article** Evans Clark, "March of the Machine Makes Idle Hands," *New York Times*, February 26, 1928.

xxi **Marvin Minsky, the MIT researcher** Brad Darrach, "Meet Shaky, the First Electronic Person," *Life*, November 20, 1970.

xxi **Oxford University study** Carl Benedikt Frey and Michael A. Osborne, "The Future of Employment: How Susceptible Are Jobs to Computerisation?," Oxford Martin Programme on Technology and Employment, September 17, 2013

xxi **By 2017, three in four American adults believed** Gallup and Northeastern University, "Optimism and Anxiety: Views on the Impact of Artificial Intelligence and Higher Education's Response," 2017.

xxii **Tyson Foods, the meat producer** Jacob Bunge and Jesse Newman, "Tyson Turns to Robot Butchers, Spurred by Coronavirus Outbreaks," *Wall Street Journal*, July 10, 2020.

xxii **FedEx started using package-sorting robots** Christopher Mims, "As E-Commerce Booms, Robots Pick Up Human Slack," *Wall Street Journal*, August 8, 2020.

xxii **Shopping centers, apartment complexes, and grocery stores** Michael Corkery and David Gelles, "Robots Welcome to Take Over, as Pandemic Accelerates Automation," *New York Times*, April 10, 2020.

xxii **McKinsey, the giant consulting firm** Chris Bradley, Martin Hirt, Sara Hudson, Nicholas Northcote, and Sven Smit, "The Great Acceleration," McKinsey, July 14, 2020.

xxii **Microsoft CEO Satya Nadella claimed** Jared Spataro, "2 Years of Digital Transformation in 2 Months," Microsoft 365 (blog), April 30, 2020.

xxii **In March 2020, a survey by the accounting firm EY** PA Media, "Bosses Speed Up Automation as Virus Keeps Workers Home," *The Guardian*, March 29, 2020.

xxii **David Autor, an MIT economist** Peter Dizikes, "The Changing World of Work," *MIT News*, May 18, 2020.

Part I The Machines

One Birth of a Suboptimist

8 **writes one such optimist** Byron Reese, *The Fourth Age* (New York: Atria Books, 2018).

8 ***Wired* declared in a 2020 article** Will Knight, "AI Is Coming for Your Most Mind-Numbing Office Tasks," *Wired*, March 14, 2020.

13 **workers routinely faced brutal conditions** Emma Griffin, *Liberty's Dawn: A People's History of the Industrial Revolution* (New Haven: Yale University Press, 2013).

14 **After the onset of the Industrial Revolution** Gregory Clark, "The Condition of the Working-Class in England, 1209–2003," *Journal of Political Economy* (2005).

14 **This gap, which was described by Friedrich Engels** Robert C. Allen, "Engels' Pause: Technical Change, Capital Accumulation, and Inequality," *Explorations in Economic History* (2008).

14 **In particular, two economists** Daron Acemoglu and Pascual Restrepo, "Automation and New Tasks: How Technology Displaces and Reinstates Labor," *Journal of Economic Perspectives* (2019).

15 **A 2019 McKinsey report** Kelemwork Cook, Duwain Pinder, Shelley Stewart, Amaka Uchegbu, and Jason Wright, "The Future of Work in Black America," McKinsey, October 4, 2019.

15 **Oxford University economist Carl Benedikt Frey** Carl Benedikt Frey, *The Technology Trap: Capital, Labor, and Power in the Age of Automation* (Princeton, N.J.: Princeton University Press, 2019).

16 **Overall rates of depression and anxiety** Jean M. Twenge, "Are

Mental Health Issues on the Rise?," *Psychology Today,* October 12, 2015.

17 **The historian David Nye writes** David E. Nye, *Electrifying America: Social Meaning of a New Technology* (Cambridge, Mass.: MIT Press, 1990).

17 **Mary L. Gray and Siddharth Suri have written** Mary L. Gray and Siddharth Suri, *Ghost Work: How to Stop Silicon Valley from Building a New Global Underclass* (New York: Houghton Mifflin Harcourt, 2019).

18 **In China, "data labeling" companies** Li Yuan, "How Cheap Labor Drives China's A.I. Ambitions," *New York Times,* November 25, 2018.

19 **One study, a 2019 preprint meta-analysis** Gagan Bansal et al., "Does the Whole Exceed Its Parts? The Effect of AI Explanations on Complementary Team Performance," *ArXiv,* June 2020.

20 **A 2014 study led by researchers at the University of Buffalo** Kenneth W. Regan et al., "Human and Computer Preferences at Chess," MPREF@AAAI, 2014.

20 **Accenture, the consulting firm, surveyed one thousand** H. James Wilson, Paul R. Daugherty, and Nicola Morini-Bianzino, "The Jobs That Artificial Intelligence Will Create," *MIT Sloan Management Review,* Summer 2017.

21 **Cognizant, a rival consulting firm, recently released a list** Benjamin Pring et al., "21 Jobs of the Future: A Guide to Getting— and Staying—Employed for the Next 10 Years," Cognizant, 2017.

Two The Myth of the Robot-Proof Job

25 **In 1895, Lord Kelvin, a well-known British physicist** Michael Marshall, "10 Impossibilities Conquered by Science," *New Scientist,* April 3, 2008.

25 **In 1962, Yehoshua Bar-Hillel, an Israeli mathematician** C.I.J.M. Stuart, *Report of the Fifteenth Annual (First International) Round Table Meeting on Linguistics and Language Studies* (Washington, D.C.: Georgetown University Press, 1964).

26 **as of 2018, Google Translate was processing** Corbin Davenport, "Google Translate Processes 143 Billion Words Every Day," Android Police, October 9, 2018.

26 **My favorite bad machine prediction** "Airport Ticket Machines Gain," *New York Times,* July 9, 1984.

26 **A 2014 study by Oxford University researchers** Stuart Armstrong, Kaj Sotala, and Séan S. ÓhÉigeartaigh, "The Errors, Insights, and Lessons of Famous AI Predictions—and What They Mean for the Future," *Journal of Experimental & Theoretical Artificial Intelligence* (2014).

27 **Richard and Daniel Susskind interviewed professionals** Richard E. Susskind and Daniel Susskind, *The Future of the Professions: How Technology Will Transform the Work of Human Experts* (Oxford: Oxford University Press, 2015).

27 **A 2017 Gallup survey found** Gallup and Northeastern University, "Optimism and Anxiety: Views on the Impact of Artificial Intelligence and Higher Education's Response," 2017.

27 **In 2019, Wendy MacNaughton, a journalist and illustrator** Wendy MacNaughton, "What Truck Drivers Think About Autonomous Trucking," *New York Times,* May 30, 2019.

28 **A 2019 study by the Brookings Institution** Mark Muro, Jacob Whiton, and Robert Maxim, "What Jobs Are Affected by AI? Better-Paid, Better-Educated Workers Face the Most Exposure," Brookings Institution, November 20, 2019.

29 **In 2017, JPMorgan Chase began using a software program** Hugh Son, "JPMorgan Software Does in Seconds What Took Lawyers 360,000 Hours," Bloomberg, February 27, 2017.

29 **Many top financial firms use Kensho** Nathaniel Popper, "The Robots Are Coming for Wall Street," *New York Times Magazine,* February 25, 2016.

29 **A 2019 report by Wells Fargo** Alfred Liu, "Robots to Cut 200,000 U.S. Bank Jobs in Next Decade, Study Says," Bloomberg, October 1, 2019.

29 **In 2018, a Chinese tech company built** Laura Yan, "Chinese AI Beats Doctors in Diagnosing Brain Tumors," *Popular Mechanics,* July 14, 2018.

29 **The same year, American researchers** Jameson Merkow et al., "DeepRadiologyNet: Radiologist Level Pathology Detection in CT Head Images," *ArXiv* preprint (2017).

30 **In a 2018 study, twenty top U.S. corporate lawyers** Jonathan Marciano, "20 Top Lawyers Were Beaten by Legal AI. Here Are Their Surprising Responses," *Hacker Noon,* October 25, 2018.

30 **In 2017, Google released AutoML** Tom Simonite, "Google's AI Experts Try to Automate Themselves," *Wired,* April 16, 2019.

31 **"overall, it took less time"** GPT-3, "A Robot Wrote This Entire
   Article. Are You Scared Yet, Human?," *The Guardian*, September 8,
   2020.

31 **Stanford researchers recently developed Woebot** Megan
   Molteni, "The Chatbot Therapist Will See You Now," *Wired*, June 7,
   2017.

31 **Early research on the effectiveness of eldercare robots** Mikaela
   Law et al., "Developing Assistive Robots for People with Mild
   Cognitive Impairment and Mild Dementia: A Qualitative Study with
   Older Adults and Experts in Aged Care," *BMJ Open* (2019).

32 **A 2019 study led by Eva G. Krumhuber** Eva G. Krumhuber et
   al., "Emotion Recognition from Posed and Spontaneous Dynamic
   Expressions: Human Observers Versus Machine Analysis," *Emotion*,
   2019.

33 **The journalist Clive Thompson recently wrote about Juke-
   deck** Clive Thompson, "What Will Happen When Machines Write
   Songs Just as Well as Your Favorite Musician?," *Mother Jones*, March/
   April 2019.

35 **In 2017, an Amazon research team developed** Thuy Ong,
   "Amazon's New Algorithm Designs Clothing by Analyzing a Bunch of
   Pictures," *The Verge*, August 14, 2017.

35 **Glitch, an AI fashion company started by two MIT graduates**
   Rob Dozier, "This Clothing Line Was Designed by AI," *Vice*, June 3,
   2019.

Three How Machines Really Replace Us

37 **Walmart brought in a fleet of floor-cleaning robots** Drew
   Harwell, "As Walmart Turns to Robots, It's the Human Workers Who
   Feel Like Machines," *Washington Post*, June 6, 2019.

39 **the technology writer Brian Merchant** Brian Merchant, "There's
   an Automation Crisis Underway Right Now, It's Just Mostly Invisible," *Gizmodo*, October 11, 2019.

39 **In their book *Competing in the Age of AI*** Marco Iansiti and
   Karim R. Lakhani, *Competing in the Age of AI* (Boston: Harvard
   Business Review Press, 2020).

Four The Algorithmic Manager

44 **"I felt so stifled"** David Noble, *Forces of Production: A Social
   History of Industrial Automation* (New York: Knopf, 1984).

44 **Every weekday, Conor Sprouls** Kevin Roose, "A Machine May Not Take Your Job, but One Could Become Your Boss," *New York Times,* June 23, 2019.

45 **Amazon uses complex algorithms to track the productivity** Colin Lecher, "How Amazon Automatically Tracks and Fires Warehouse Workers for 'Productivity,'" *The Verge,* April 25, 2019.

45 **IBM has used Watson, its AI platform** Tristan Greene, "IBM Is Using Its AI to Predict How Employees Will Perform," *TheNextWeb,* July 10, 2018.

46 **Percolata, a Silicon Valley start-up** Hazel Sheffield, "The Great Data Leap: How AI Will Transform Recruitment and HR," *Financial Times,* November 4, 2019.

47 **The company, whose clients include Sweetgreen and OfferUp** Daisuke Wakabayashi, "Firm Led by Google Veterans Uses AI to 'Nudge' Workers Toward Happiness," *New York Times,* December 31, 2018.

48 **a setting in Instacart's app that applied customer tips** Kevin Roose, "After Uproar, Instacart Backs Off Controversial Tipping Policy," *New York Times,* February 6, 2019.

48 **A 2019 study of Uber drivers** Mareike Möhlmann and Ola Henfridsson, "What People Hate About Being Managed by Algorithms, According to a Study of Uber Drivers," *Harvard Business Review,* August 30, 2019.

Five Beware of Boring Bots

50 **"Your actions indicate you intentionally misled"** Bauserman v. Unemployment Ins. Agency, Case No. 333181 (Michigan Supreme Court, 2018).

51 **Virginia Eubanks, a professor of political science** Virginia Eubanks, *Automating Inequality: How High-Tech Tools Profile, Police, and Punish the Poor* (New York: St. Martin's Press, 2018).

52 **In 2007, a glitch in an automated system** "Computer Glitch May Have Cost Thousands Their Benefits," *Orange County Register,* March 2, 2007.

52 **In Ohio, a yearslong project** Rita Price, "New Computer System Causing Confusion, Benefit Delays for Ohio Food-stamp Recipients," *Columbus Dispatch,* January 21, 2019.

52 **In Idaho, a flawed automated process** Colin Lecher, "What

Happens When an Algorithm Cuts Your Healthcare," *The Verge*, March 21, 2018.

53 **In 2019, Microsoft announced** James Phillips, "Announcing RPA, Enhanced Security, No-Code Virtual Agents, and More for Microsoft Power Platform," Microsoft Dynamics 365 (blog), November 4, 2019.

54 **Craig Le Clair, an analyst at Forrester Research** Craig Le Clair, *Invisible Robots in the Quiet of the Night: How AI and Automation Will Restructure the Workforce* (Forrester, 2019).

56 **MIT's Daron Acemoglu and Boston University's Pascual Restrepo** Daron Acemoglu and Pascual Restrepo, "Automation and New Tasks: How Technology Displaces and Reinstates Labor," *Journal of Economic Perspectives* (2019).

Part II The Rules

Rule 1 Be Surprising, Social, and Scarce

61 **Lovett was a working-class kid** William Lovett, *Life and Struggles of William Lovett, in His Pursuit of Bread, Knowledge, and Freedom* (Knopf, 1876).

63 **The makers of one NLG app, Wordsmith** Lance Ulanoff, "Need to Write 5 Million Stories a Week? Robot Reporters to the Rescue," *Mashable*, July 1, 2014.

63 **Another app** Steve Lohr, "In Case You Wondered, a Real Human Wrote This Column," *New York Times*, September 10, 2011.

64 **Marc Andreessen, the venture capitalist and Netscape cofounder** Hannah Kuchler, "How Silicon Valley Learnt to Love the Liberal Arts," *Financial Times Magazine*, October 31, 2017.

64 **Vinod Khosla, the venture capitalist** Vinod Khosla, "Is Majoring in Liberal Arts a Mistake for Students?," *Medium*, February 10, 2016.

65 **Even President Obama argued** Scott Jaschik, "Obama vs. Art History," *Inside Higher Education*, January 21, 2014.

67 **In a 2018 experiment, a group of AI researchers** Kevin Hartnett, "Machine Learning Confronts the Elephant in the Room," *Quanta Magazine*, September 20, 2018.

72 **Maria Popova, the creator of the Brain Pickings blog** Maria Popova, "Networked Knowledge and Combinatorial Creativity," Brain Pickings, August 1, 2011.

78 **Education, Lovett wrote** William Lovett and John Collins, *Chartism: A New Organization of the People* (London: J. Watson, 1840).

Rule 2 Resist Machine Drift

79 **"The main business of humanity"** From *Player Piano* (New York: Scribner, 1952).

83 **Doug Terry, a junior researcher at the lab** Douglas B. Terry, "A Tour Through Tapestry," *Proceedings of the 1993 ACM Conference on Organizational Computing Systems* (1993).

85 **Michael Schrage, an MIT research fellow** Michael Schrage, *Recommendation Engines* (Boston: MIT Press, 2020).

86 **YouTube has said that recommendations** Paresh Dave, "YouTube Sharpens How It Recommends Videos Despite Fears of Isolating Users," Reuters, November 28, 2017.

86 **It has been estimated that 30 percent of Amazon page views** Amit Sharma, Jake M. Hofman, and Duncan J. Watts, "Estimating the Causal Impact of Recommendation Systems from Observational Data," *Proceedings of the 2015 ACM Conference on Economics and Computation* (2015).

86 **Spotify's algorithmically generated Discover Weekly playlists** Devindra Hardawar, "Spotify's Discover Weekly Playlists Have 40 Million Listeners," *Engadget,* May 25, 2016.

86 **Netflix has said that 80 percent of the movies viewed** Ashley Rodriguez, "'Because You Watched': Netflix Finally Explains Why It Recommends Titles That Seem to Have Nothing in Common," *Quartz,* August 22, 2017.

86 **a 2018 study led by Gediminas Adomavicius** Gediminas Adomavicius, Jesse C. Bockstedt, Shawn P. Curley, and Jingjing Zhang, "Effects of Online Recommendations on Consumers' Willingness to Pay," *Information Systems Research* (2017).

88 **The technology scholar Christian Sandvig** Christian Sandvig, "Corrupt Personalization," *Social Media Collective,* June 26, 2014.

89 **Rachel Schutt, a data scientist** Steve Lohr, "Sure, Big Data Is Great. But So Is Intuition," *New York Times,* December 29, 2012.

89 **A former product manager at Facebook** Alex Kantrowitz, "Facebook Is Still Prioritizing Scale over Safety," *BuzzFeed News,* December 17, 2019.

89 **The French researcher Camille Roth divides** Camille Roth,

"Algorithmic Distortion of Informational Landscapes," *Intellectica* (2019).

90 **Amazon engineer Brent Smith** Brent Smith and Greg Linden, "Two Decades of Recommender Systems at Amazon.com," *IEEE Computer Society* (2017).

91 **Brenden Mulligan, a tech entrepreneur** Brenden Mulligan, "Reduce Friction, Increase Happiness," *TechCrunch*, October 16, 2011.

91 **Facebook CEO Mark Zuckerberg announced** Brittany Darwell, "Facebook's Frictionless Sharing Mistake," *Adweek*, January 22, 2013.

92 **Jeff Bezos, Amazon's founder** Jeff Bezos, "2018 Letter to Shareholders," Amazon.com, 2018.

92 **Uber drivers lost out on millions of dollars** Arik Jenkins, "Why Uber Doesn't Want a Built-In Tipping Option," *Fortune*, April 18, 2017.

93 **the technology critic Tim Wu calls** Tim Wu, "The Tyranny of Convenience," *New York Times*, February 16, 2018.

Rule 3 Demote Your Devices

102 **Adam Smith warned** Adam Smith, *The Wealth of Nations* (1776).

103 **The psychologist Sherry Turkle** Sherry Turkle, *Reclaiming Conversation: The Power of Talk in a Digital Age* (New York: Penguin, 2015).

103 **One such study, conducted at the University of British Columbia** Ryan J. Dwyer, Kostadin Kushlev, and Elizabeth W. Dunn, "Smartphone Use Undermines Enjoyment of Face-to-Face Social Interactions," *Journal of Experimental Social Psychology* (September 2018).

104 **using Facebook passively . . . has been shown to increase anxiety and decrease happiness** Philippe Verduyn et al., "Passive Facebook Usage Undermines Affective Well-Being: Experimental and Longitudinal Evidence," *Journal of Experimental Psychology* (2015).

104 **using Facebook more actively . . . has been shown to have more positive effects** Moira Burke and Robert E. Kraut, "The Relationship Between Facebook Use and Well-Being Depends on Communication Type and Tie Strength," *Journal of Computer-Mediated Communication* (2015).

105 **Her name was Catherine Price** Kevin Roose, "Do Not Disturb: How I Ditched My Phone and Unbroke My Brain," *New York Times*, February 23, 2019.

107 **Research has shown that being alone with our thoughts** Timothy D. Wilson et al., "Just Think: The Challenges of the Disengaged Mind," *Science* (2014).

112 **Jenny Odell . . . writes** Jenny Odell, *How to Do Nothing: Resisting the Attention Economy* (New York: Melville House, 2019).

Rule 4 Leave Handprints

115 **Mitsuru Kawai should have been panicking** Shusuke Murai, "Hands-on Toyota Exec Passes Down Monozukuri Spirit," *Japan Times*, April 15, 2018.

116 **A 1961 article in *Time* magazine** "The Automation Jobless," *Time*, February 24, 1961.

116 **Another article called factory automation** Rick Wartzman, "The First Time America Freaked Out over Automation," *Politico*, May 30, 2017.

118 **Today, Kawai is a living legend at Toyota** "Toyota's 'Oyaji' Kawai Calls to Protect Monozukuri," *Toyota News*, June 17, 2020.

119 **a former steelworker named Frederick Winslow Taylor** Frederick Winslow Taylor, *The Principles of Scientific Management* (New York: Harper & Brothers, 1915).

119 **Gary Vaynerchuk, a marketing guru and social media influencer** Ted Fraser, "I Spent a Week Living Like Gary Vaynerchuk," *Vice*, December 17, 2018.

119 **Elon Musk, the Tesla and SpaceX founder, famously works** Catherine Clifford, "Elon Musk on Working 120 Hours in a Week: 'However Hard It Was for [the Team], I Would Make It Worse for Me,'" CNBC, December 10, 2018.

120 **Marissa Mayer, the former chief executive at Yahoo** Max Chafkin, "Yahoo's Marissa Mayer on Selling a Company While Trying to Turn It Around," *Bloomberg Businessweek*, August 4, 2016.

120 **the writer Derek Thompson has called "workism"** Derek Thompson, "Workism Is Making Americans Miserable," *The Atlantic*, February 24, 2019.

121 **Yann LeCun, then the head of Facebook's AI research division** "Yann LeCun—Power & Limits of Deep Learning," accessed on YouTube, October 4, 2020.

123 **Social scientists call this the "effort heuristic"** Derrick Wirtz, Justin Kruger, William Altermatt, and Leaf Van Boven, "The Effort Heuristic," *Journal of Experimental Social Psychology* (2004).

123 **In his book *The Power of Human*** Adam Waytz, *The Power of Human: How Our Shared Humanity Can Help Us Create a Better World* (New York: W. W. Norton, 2019).

123 **One such experiment, led by Kurt Gray** Kurt Gray, "The Power of Good Intentions: Perceived Benevolence Soothes Pain, Increases Pleasure, and Improves Taste," *Social Psychological and Personality Science* (2012).

123 **the rise of craft breweries, farm-to-table restaurants, and artisanal Etsy shops** Timothy B. Lee, "Automation Is Making Human Labor More Valuable Than Ever," *Vox*, September 26, 2016.

124 **In the early days of Facebook** Glenn Fleishman, "How Facebook Devalued the Birthday," *Fast Company*, April 6, 2018.

126 **business school professors B. Joseph Pine II and James H. Gilmore** B. Joseph Pine II and James H. Gilmore, *The Experience Economy: Competing for Customer Time, Attention, and Money*, revised edition (Boston: Harvard Business Review Press, 2019).

127 **Best Buy learned this lesson out of necessity** Kevin Roose, "Best Buy's Secrets for Thriving in the Amazon Age," *New York Times*, September 18, 2017.

129 **what happened to Heath Ceramics** Hannah Wallace, "This Ceramics Company Had a Cult Following but No Money. Then, 2 New Owners Brought It Back from the Brink," *Inc.*, July/August 2019.

Rule 5 Don't Be an Endpoint

131 **"Right now, I feel like they need us"** Khristopher J. Brooks, "Why Automation Could Hit Black Workers Harder Than Other Groups," CBS News, October 10, 2019.

131 **The product was a voice-based AI assistant called Duplex** Chris Welch, "Google Just Gave a Stunning Demo of Assistant Making an Actual Phone Call," *The Verge*, May 8, 2018.

132 **"Google Duplex is the most incredible"** Tweet by @chrismessina, May 8, 2018.

135 **As the journalist Martin Ford writes** Martin Ford, *Rise of the Robots: Technology and the Threat of Mass Unemployment* (London: OneWorld Publications, 2015).

135 **a 2018 New Yorker piece** Atul Gawande, "Why Doctors Hate Their Computers," *The New Yorker*, November 12, 2018.

136 **Emily Silverman, a physician in San Francisco** Emily Silverman, "Our Hospital's New Software Frets About My 'Deficiencies,'" *New York Times*, November 1, 2019.

136 **When implemented properly** Catherine M. DesRoches et al., "Electronic Health Records in Ambulatory Care—A National Survey of Physicians," *New England Journal of Medicine* (2008).

136 **the Lordstown factory was a dazzling tribute to modernity** Gwynn Guilford, "GM's Decline Truly Began with Its Quest to Turn People into Machines," *Quartz*, December 30, 2018.

137 **one Lordstown worker described his daily routine** Peter Herman, *In the Heart of the Heart of the Country: The Strike at Lordstown* (Greenwich, Conn.: Fawcett, 1975).

137 **formed "humanization teams" tasked with improving conditions** Bennett Kremen, "Lordstown—Searching for a Better Way of Work," *New York Times*, September 9, 1973.

137 **Edward Cole, then GM's president, said in a speech** Agis Salpukas, "Workers Increasingly Rebel Against Boredom on Assembly Line," *New York Times*, April 2, 1972.

138 **A July 2020 survey by Gartner** "Gartner Survey Reveals 82% of Company Leaders Plan to Allow Employees to Work Remotely Some of the Time," Gartner, July 14, 2020.

139 **Adobe CEO Shantanu Narayen complained** Kevin Stankiewicz, "Adobe CEO Says Offices Provide Some Boost to Productivity That Remote Work Lacks," CNBC, August 11, 2020.

139 **Reed Hastings, the chief executive of Netflix, called remote work a "pure negative"** Joe Flint, "Netflix's Reed Hastings Deems Remote Work 'a Pure Negative,'" *Wall Street Journal*, September 7, 2020.

139 **Studies have found that groups of people located** Jerry Useem, "When Working from Home Doesn't Work," *The Atlantic*, November 2017.

139 **co-authors of academic papers who are located closer** Kyung-joon Lee, John S. Brownstein, Richard G. Mills, and Isaac S. Kohane, "Does Collocation Inform the Impact of Collaboration?," *PLoS ONE* (2010).

139 **team cohesion suffers** Tammy D. Allen, Timothy D. Golden, and Kristen M. Shockley, "How Effective Is Telecommuting? Assessing the Status of Our Scientific Findings," *Psychological Science in the Public Interest* (2015).

140 **John Sullivan, a management professor** Steve Henn, "'Serendipitous Interaction' Key to Tech Firms' Workplace Design," NPR, March 13, 2013.

141 **At GitLab, an open-source collaboration platform** Sid Sijbrandij, "'Virtual Coffee Breaks' Encourage Remote Workers to Interact Like They Would in an Office," *Quartz*, December 6, 2017.

141 **At the Seattle-based software company Seeq** Ben Johnson, "How Well Do You Really Know Your Coworkers? A Virtual Company Shares All," Seeq Culture Blog, May 15, 2018.

141 **Automattic, the all-remote maker of WordPress, organizes** Matt Mullenweg, "The Importance of Meeting In-Person," Unlucky in Cards (blog), October 16, 2018.

Rule 6 Treat AI Like a Chimp Army

144 **"We fixed a technical issue"** Mariel Padilla, "Facebook Apologizes for Vulgar Translation of Chinese Leader's Name," *New York Times,* January 18, 2020.

146 **a customer browsing Amazon noticed some disturbing shirts** Eric Limer, "Amazon Blocks the Sale of Gross, Auto-Generated 'Keep Calm and Rape Her' Shirts," *Gizmodo,* March 2, 2013.

148 **In a recent article in the *MIT Sloan Management Review*** Barry Libert, Megan Beck, and Thomas H. Davenport, "Self-Driving Companies Are Coming," August 29, 2019.

150 **Shane, who expanded AI Weirdness into a book** Janelle Shane, *You Look Like a Thing and I Love You* (New York: Headline, 2019).

150 **A trading firm called Knight Capital, for example** Nathaniel Popper, "Knight Capital Says Trading Glitch Cost It $440 Million," *New York Times,* August 2, 2012.

151 **In 2018, internal tests obtained** Casey Ross and Ike Swetlitz, "IBM's Watson Supercomputer Recommended 'Unsafe and Incorrect' Cancer Treatments, Internal Documents Show," Stat, July 25, 2018.

151 **decades of systematic overpolicing** Rashida Richardson, Jason M. Schultz, and Kate Crawford, "Dirty Data, Bad Predictions: How Civil Rights Violations Impact Police Data, Predictive Policing Systems, and Justice," *New York University Law Review,* Online Feature (2019).

152 **A 2016 investigation by ProPublica** Julia Angwin, Jeff Larson, Surya Mattu, and Lauren Kirchner, "Machine Bias," ProPublica, May 23, 2016.

152 **a 2018 interview with the journalist Martin Ford** Martin Ford, *Architects of Intelligence: The Truth About AI from the People Building It* (Birmingham, U.K.: Packt Publishing, 2018).

153 **Musk stopped the belts** Dana Hull, "Musk Says Excessive Automation Was 'My Mistake,'" Bloomberg, April 13, 2018.

153 **In their book *Turning Point*** John R. Allen and Darrell M. West, *Turning Point: Policymaking in the Era of Artificial Intelligence* (Washington, D.C.: Brookings Institution Press, 2020).

154 **In 2019, Senators Cory Booker and Ron Wyden** "Booker, Wyden, Clarke Introduce Bill Requiring Companies to Target Bias in Corporate Algorithms," Senator Booker's official site, April 10, 2019.

155 **the Chicago Police Department announced** "Chicago Police Drop Clearview Facial Recognition Technology," Associated Press, May 29, 2020.

155 **There are now "AI auditors"** Erin Winick, "This Company Audits Algorithms to See How Biased They Are," *MIT Technology Review,* May 9, 2018.

Rule 7   Build Big Nets and Small Webs

158 **A decade ago, this lot would have been full** Kevin Roose, "The Life, Death, and Rebirth of BlackBerry's Hometown," *Fusion,* February 8, 2015.

161 **a widespread labor practice called *shukko*** Frederik L. Schodt, *Inside the Robot Kingdom: Japan, Mechatronics, and the Coming Robotopia* (New York: Harper & Row, 1988).

161 **In Sweden, workers who lose their jobs to automation** Peter S. Goodman, "The Robots Are Coming, and Sweden Is Fine," *New York Times,* December 27, 2017.

162 **Some leaders, including Bill Gates and New York City mayor Bill DeBlasio** Richard Rubin, "The Robot Tax Debate Heats Up," *Wall Street Journal,* January 8, 2020.

162 **A 2019 report by the World Economic Forum** "Towards a Reskilling Revolution: Industry-Led Action for the Future of Work," World Economic Forum, January 22, 2019.

163 **Airbnb . . . was forced to lay off 25 percent of its staff** Erin Griffith, "Airbnb Was Like a Family. Until the Layoffs Started," *New York Times,* July 17, 2020.

164 **Executives from Accenture** Sarah Fielding, "Accenture and

Verizon Lead Collaborative Effort to Help Furloughed or Laid-Off Workers Find a New Job," *Fortune,* April 14, 2020.

164 **According to the historian David E. Nye** David E. Nye, *Electrifying America: Social Meaning of a New Technology* (Cambridge, Mass.: MIT Press, 1990).

Rule 8 Learn Machine-Age Humanities

167 **Daniel Goleman, the psychologist** Daniel Goleman, *Focus: The Hidden Driver of Excellence* (New York: A&C Black, 2013).

168 **periods of meditation even as short as eight minutes** Mengran Xu et al., "Mindfulness and Mind Wandering: The Protective Effects of Brief Meditation in Anxious Individuals," *Consciousness and Cognition* (2017).

168 **as the historian and author Yuval Noah Harari writes** Yuval Noah Harari, *21 Lessons for the 21st Century* (New York: Spiegel & Grau, 2018).

170 **It's run by Tricia Hersey** "Listen: You Are Worthy of Sleep," *Social Distance* podcast, April 30, 2020.

171 **Studies conducted by neuroscientists at the Walter Reed Army Institute of Research** William D. S. Killgore et al., "The Effects of 53 Hours of Sleep Deprivation on Moral Judgment," *Sleep* (2007).

171 **lowers our emotional intelligence** William D. S. Killgore et al., "Sleep Deprivation Reduces Perceived Emotional Intelligence and Constructive Thinking Skills," *Sleep Medicine* (2007).

171 **harms our interpersonal communication skills** Yvonne Harrison and James A. Horne, "Sleep Deprivation Affects Speech," *Sleep* (2010).

171 **In Japan, a 2019 law limited worker overtime** Aki Tanaka and Trent Sutton, "Significant Changes to Japan's Labor Laws Will Take Effect in April 2019: Are You Prepared?," *Littler,* February 12, 2019.

172 **A French law that went into effect in 2017** Alanna Petroff and Océane Cornevin, "France Gives Workers 'Right to Disconnect' from Office Email," CNN, January 2, 2017.

172 **At Harvard, incoming freshmen** John Michael Baglione, "Countering College's Culture of Sleeplessness," *Harvard Gazette,* August 24, 2018.

173 **In 2015, a group of Stanford researchers** Sarah McGrew et al.,

"Can Students Evaluate Online Sources? Learning from Assessments of Civic Online Reasoning," *Theory & Research in Social Education* (2018).

174 **In fact, one study found** Niraj Chokshi, "Older People Shared Fake News on Facebook More Than Others in 2016 Race, Study Says," *New York Times,* January 10, 2019.

174 **In a 2018 report for the nonprofit organization** Monica Bulger and Patrick Davison, "The Promises, Challenges, and Futures of Media Literacy," *Journal of Media Literacy Education* (2018).

175 **Chen believes that these skills** Frank Chen, "Humanity + AI: Better Together," Andreessen Horowitz (blog), February 22, 2019.

176 **One 2015 study that tracked children** Damon E. Jones, Mark Greenberg, and Max Crowley, "Early Social-Emotional Functioning and Public Health: The Relationship Between Kindergarten Social Competence and Future Wellness," *American Journal of Public Health* (2015).

176 **Another study in 2017 found that kids who participated** Rebecca D. Taylor, Eva Oberle, Joseph A. Durlak, and Roger P. Weissberg, "Promoting Positive Youth Development Through School-Based Social and Emotional Learning Interventions: A Meta-Analysis of Follow-Up Effects," *Child Development* (2017).

178 **Jack Dorsey, the chief executive of Twitter** *The Daily* podcast, "Jack Dorsey on Twitter's Mistakes," *New York Times,* August 7, 2020.

178 **In Canada, when you graduate from engineering school** Erin Hudson, "An Inside Look at the 'Not Secretive but Modestly Discrete' Iron Ring Ritual for Canadian Trained-Engineers," *The Sheaf,* January 10, 2013.

Rule 9   Arm the Rebels

181 **"We are in great haste to construct a magnetic telegraph"** Henry David Thoreau, *Walden, Civil Disobedience, and Other Writings* (New York: W. W. Norton, 2008).

181 **Bagley lived in Lowell, Massachusetts** Cara Giaimo, "Sarah Bagley, the Voice of America's Early Women's Labor Movement," *Atlas Obscura,* March 8, 2017.

182 **She lampooned the factory owners** Philip Dray, *There Is Power in a Union: The Epic Story of Labor in America* (New York: Anchor Books, 2011).

184 **underappreciated heroes like Katherine Johnson, Dorothy**

Vaughan, and Mary Jackson Margot Lee Shetterly, *Hidden Figures: The American Dream and the Untold Story of the Black Women Mathematicians Who Helped Win the Space Race* (New York: William Morrow, 2016).

184 **These are people like Jazmyn Latimer** Vanessa Taylor, "This Founder Is Using Technology to Clear Criminal Records," *Afrotech,* February 22, 2019.

185 **Or Rohan Pavuluri** Kevin Roose, "The 2018 Good Tech Awards," *New York Times,* December 21, 2018.

185 **Or Joy Buolamwini and Timnit Gebru** Kevin Roose, "The 2019 Good Tech Awards," *New York Times,* December 30, 2019.

185 **Or Sasha Costanza-Chock** Sasha Costanza-Chock, *Design Justice: Community-Led Practices to Build the Worlds We Need* (Boston: MIT Press, 2020).

186 **a term coined by the evolutionary biologist Stuart Kauffman** Stuart Kauffman, *The Origins of Order: Self-Organization and Selection in Evolution* (New York: Oxford University Press, 1993).

188 **Sarah Bagley made history again** Madeleine B. Stern, *We the Women: Career Firsts of Nineteenth-Century America* (Lincoln, Neb.: Bison Books, 1994).

About the Author

KEVIN ROOSE is a technology columnist for *The New York Times*. He is the host of the *Rabbit Hole* podcast and a regular guest on *The Daily*. He writes and speaks regularly about topics including automation and AI, social media, disinformation and cybersecurity, and digital wellness. Previously, he was a writer at *New York* magazine and the co–executive producer of *Real Future,* a TV documentary series about technology. He is the *New York Times* bestselling author of two previous books, *Young Money* and *The Unlikely Disciple*. He lives in Oakland, California.

About the Type

This book was set in Fairfield, the first typeface from the hand of the distinguished American artist and engraver Rudolph Ruzicka (1883–1978). Ruzicka was born in Bohemia (in the present-day Czech Republic) and came to America in 1894. He set up his own shop, devoted to wood engraving and printing, in New York in 1913 after a varied career working as a wood engraver, in photoengraving and banknote printing plants, and as an art director and freelance artist. He designed and illustrated many books, and was the creator of a considerable list of individual prints—wood engravings, line engravings on copper, and aquatints.